Collins

GCSE 9-1
Chemistry
in a week

Dan Evans

Revision Planner

HT indicates content that is Higher Tier only.

WS indicates 'Working Scientifically' content, which covers practical skills and data-related concepts.

Atoms and the periodic table

Atoms and elements

- All substances are made of **atoms**.
- An atom is the smallest part of an **element** that can exist.
- There are approximately 100 different elements, all of which are shown in the **periodic table**.

Below is part of the periodic table, showing the names and symbols of the first 20 elements.

Atoms of each element are represented by a chemical symbol in the periodic table.

Some elements have more than one letter in their symbol. The first letter is always a capital and the other letter is lower case.

Chemical equations

We can use word or symbol equations to represent chemical reactions.

Word equation:

sodium + oxygen $\longrightarrow$ sodium oxide

Symbol equation:

$4Na + O_2 \longrightarrow 2Na_2O$

1 hydrogen H										2 helium He
3 lithium Li	4 beryllium Be		5 boron B	6 carbon C	7 nitrogen N	8 oxygen O	9 fluorine F	10 neon Ne		
11 sodium Na	12 magnesium Mg		13 aluminium Al	14 silicon Si	15 phosphorus P	16 sulfur S	17 chlorine Cl	18 argon Ar		
19 potassium K	20 calcium Ca									

The symbol for magnesium is **Mg**

The symbol for the element oxygen is **O**

Elements, compounds and mixtures

Element	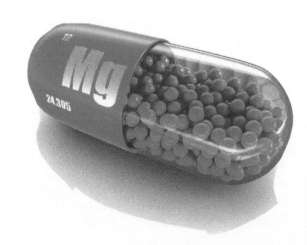	Elements contain only one type of atom and consist of single atoms or atoms bonded together
Compound		Compounds contain two (or more) elements that are chemically combined in fixed proportions
Mixture		Mixtures contain two or more elements (or compounds) that are together but not chemically combined

Separating mixtures

Compounds can only be separated by chemical reactions but **mixtures** can be separated by physical processes (not involving chemical reactions) such as filtration, crystallisation, simple (and fractional) distillation and chromatography.

How different mixtures can be separated	
Method of separation	**Diagram**
Filtration Used for separating insoluble solids from liquids, e.g. sand from water	Filter paper Filter funnel Sand and water Sand Beaker Clear water (filtrate)
Crystallisation Used to separate solids from solutions, e.g. obtaining salt from salty water	Evaporating dish Mixture Wire gauze Tripod stand Bunsen burner
Simple distillation Simple distillation is used to separate a liquid from a solution, e.g. water from salty water	Thermometer Water out Round-bottomed flask Liebig condenser Water in Heat
Fractional distillation Fractional distillation is used to separate liquids that have different boiling points, e.g. ethanol and water	
Chromatography Used to separate dyes, e.g. the different components of ink	Solvent front Separated dyes Chromatography paper Ink spots Pencil line Solvent

SUMMARY

- An atom is the smallest part of an element.
- The elements are shown in the periodic table.
- Mixtures can be separated by physical processes; compounds must be separated by chemical reactions.

QUESTIONS

QUICK TEST

1. What are all substances made from?

2. What name is given to a substance that contains atoms of different elements chemically joined together?

3. Which method of separation would you use to obtain salt from salt water?

EXAM PRACTICE

1. A student is provided with a mixture of magnesium chloride and magnesium oxide. Magnesium chloride is soluble in water but magnesium oxide is insoluble.

   a) Write a balanced symbol equation for the reaction between magnesium and oxygen to form magnesium oxide.

      Include state symbols in your answer. **[2 marks]**

   b) What is meant by the term 'mixture'? **[1 mark]**

   c) Describe how the mixture of magnesium chloride and magnesium oxide could be separated. **[3 marks]**

Atomic structure

(ws) Scientific models of the atom

Scientists had originally thought that atoms were tiny spheres that could not be divided.

John Dalton conducted experiments in the early 19th century and concluded that…

- all matter is made of indestructible atoms
- atoms of a particular element are identical
- atoms are rearranged during chemical reactions
- compounds are formed when two or more different types of atom join together.

Upon discovery of the electron by **J. J. Thomson** in 1897, the 'plum pudding' model suggested that the atom was a ball of positive charge with negative electrons embedded throughout.

The results from **Rutherford**, **Geiger** and **Marsden's** alpha scattering experiments (1911–1913) led to the plum pudding model being replaced by the nuclear model.

In this experiment, alpha particles (which are positive) are fired at a thin piece of gold. A few of the alpha particles do not pass through the gold and are deflected. Most went straight through the thin piece of gold. This led Rutherford, Geiger and Marsden to suggest that this is because the positive charge of the atom is confined in a small volume (now called the nucleus).

Niels Bohr adapted the nuclear model in 1913, by suggesting that electrons orbit the nucleus at specific distances. Bohr's theoretical calculations were backed up by experimental results.

Later experiments led to the idea that the positive charge of the nucleus was subdivided into smaller particles (now called protons), with each particle having the same amount of positive charge.

The work of **James Chadwick** suggested in 1932 that the nucleus also contained neutral particles that we now call neutrons.

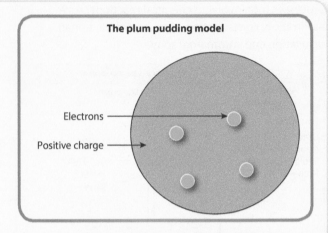

The plum pudding model

Electrons

Positive charge

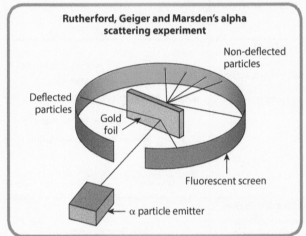

Rutherford, Geiger and Marsden's alpha scattering experiment

Non-deflected particles

Deflected particles

Gold foil

Fluorescent screen

α particle emitter

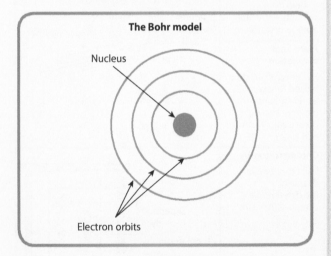

The Bohr model

Nucleus

Electron orbits

Properties of atoms

Particle	Relative charge	Relative mass
Proton	+1	1
Neutron	0	1
Electron	−1	negligible

Atoms have a neutral charge. This is because the number of protons is equal to the number of electrons.

● Atoms of different elements have different numbers of protons. This number is called the atomic number.
● Atoms are very small, having a radius of approximately 0.1 nm (1×10^{-10} m).
● The radius of the nucleus is approximately $\frac{1}{10\,000}$ of the size of the atom.

The **mass number** tells you the total number of protons and neutrons in an atom

$$^{23}_{11}\text{Na}$$

The **atomic number** tells you the number of protons in an atom

mass number	−	atomic number	=	number of neutrons

Some atoms can have different numbers of neutrons. These atoms are called **isotopes**. The existence of isotopes results in the relative atomic mass of some elements, e.g. chlorine, not being whole numbers.

> **HT** Chlorine exists as two isotopes. Chlorine-35 makes up 75% of all chlorine atoms. Chlorine-37 makes up the other 25%. We say that the abundance of chlorine-35 is 75%.
>
> The relative atomic mass of chlorine can be calculated as follows:
>
> $$\frac{(\text{mass of isotope 1} \times \text{abundance}) + (\text{mass of isotope 2} \times \text{abundance})}{100}$$
>
> $$= \frac{(35 \times 75) + (37 \times 25)}{100} = 35.5$$

SUMMARY

● Notable scientists who worked on atomic structure are John Dalton; J. J. Thompson; Rutherford, Geiger and Marsden; Niels Bohr and James Chadwick.
● Atoms have protons (positive), neutrons (neutral) and electrons (negative) and are neutral.

QUESTIONS

QUICK TEST

1. What is the relative charge of a proton?

2. How many protons, neutrons and electrons are present in the following atom?

$$^{24}_{12}\text{Mg}$$

3. What name is given to atoms of the same element which have the same number of protons but different numbers of neutrons?

EXAM PRACTICE

1. The diagram below shows J. J. Thomson's plum pudding model of an atom and Bohr's nuclear model of an atom.

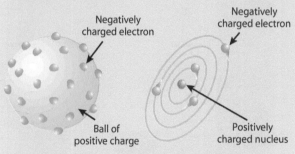

Negatively charged electron

Negatively charged electron

Ball of positive charge

Positively charged nucleus

Describe the key differences between these two different models of the atom. **[3 marks]**

Electronic structure & the periodic table

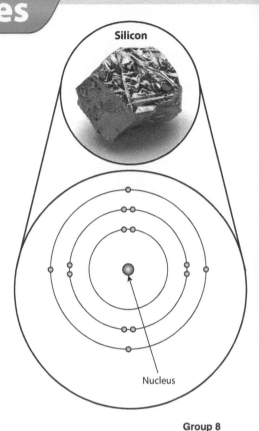

Silicon

Nucleus

Electronic structures

Electrons in an atom occupy the lowest available energy level (shell).

The first energy level (closest to the nucleus) can hold up to two electrons. The second and third energy levels can hold up to eight electrons.

For example, silicon has the atomic number 14. This means that there are 14 protons in the nucleus of a silicon atom and therefore there must be 14 electrons (so that the atom is neutral).

The electronic structure of silicon can be written as 2, 8, 4 or shown in a diagram like the one on the right.

Silicon is in group 4 of the periodic table. This is because it has four electrons in its outer shell. The chemical properties (reactions) of an element are related to the number of electrons in the outer shell of the atom.

The electronic structure of the first 20 elements is shown here.

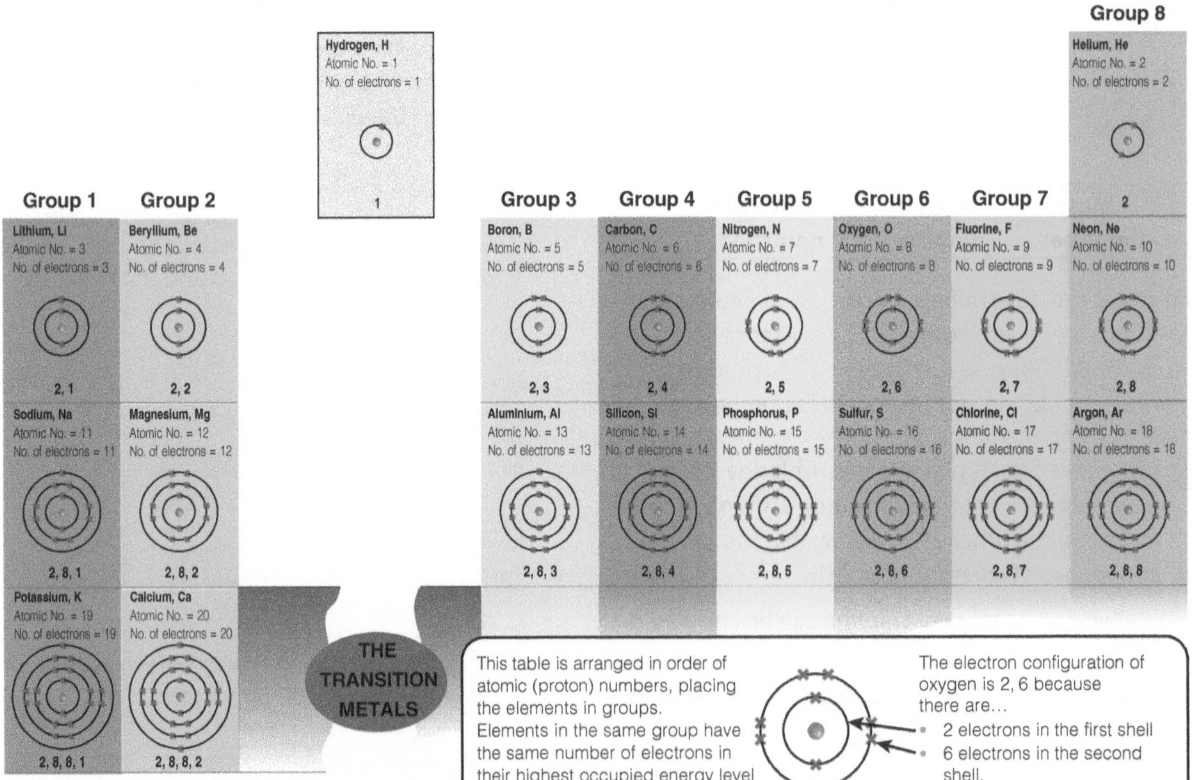

Group 8

Hydrogen, H Atomic No. = 1 No. of electrons = 1	**Helium, He** Atomic No. = 2 No. of electrons = 2
1	2

Group 1	Group 2	Group 3	Group 4	Group 5	Group 6	Group 7	
Lithium, Li Atomic No. = 3 No. of electrons = 3	**Beryllium, Be** Atomic No. = 4 No. of electrons = 4	**Boron, B** Atomic No. = 5 No. of electrons = 5	**Carbon, C** Atomic No. = 6 No. of electrons = 6	**Nitrogen, N** Atomic No. = 7 No. of electrons = 7	**Oxygen, O** Atomic No. = 8 No. of electrons = 8	**Fluorine, F** Atomic No. = 9 No. of electrons = 9	**Neon, Ne** Atomic No. = 10 No. of electrons = 10
2, 1	2, 2	2, 3	2, 4	2, 5	2, 6	2, 7	2, 8
Sodium, Na Atomic No. = 11 No. of electrons = 11	**Magnesium, Mg** Atomic No. = 12 No. of electrons = 12	**Aluminium, Al** Atomic No. = 13 No. of electrons = 13	**Silicon, Si** Atomic No. = 14 No. of electrons = 14	**Phosphorus, P** Atomic No. = 15 No. of electrons = 15	**Sulfur, S** Atomic No. = 16 No. of electrons = 16	**Chlorine, Cl** Atomic No. = 17 No. of electrons = 17	**Argon, Ar** Atomic No. = 18 No. of electrons = 18
2, 8, 1	2, 8, 2	2, 8, 3	2, 8, 4	2, 8, 5	2, 8, 6	2, 8, 7	2, 8, 8
Potassium, K Atomic No. = 19 No. of electrons = 19	**Calcium, Ca** Atomic No. = 20 No. of electrons = 20						
2, 8, 8, 1	2, 8, 8, 2						

THE TRANSITION METALS

This table is arranged in order of atomic (proton) numbers, placing the elements in groups. Elements in the same group have the same number of electrons in their highest occupied energy level (outer shell).

The electron configuration of oxygen is 2, 6 because there are…
- 2 electrons in the first shell
- 6 electrons in the second shell.

The periodic table

The elements in the periodic table are arranged in order of increasing atomic (proton) number.

The table is called a **periodic table** because similar properties occur at regular intervals.

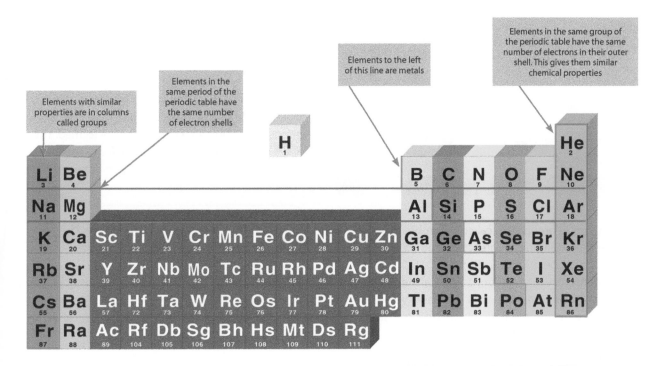

The periodic table with annotations:
- Elements with similar properties are in columns called groups
- Elements in the same period of the periodic table have the same number of electron shells
- Elements to the left of this line are metals
- Elements in the same group of the periodic table have the same number of electrons in their outer shell. This gives them similar chemical properties

Development of the periodic table

Before the discovery of protons, neutrons and electrons, early attempts to classify the elements involved placing them in order of their atomic weights. These early attempts resulted in incomplete tables and the placing of some elements in appropriate groups based on their chemical properties.

Dmitri Mendeleev overcame some of these problems by leaving gaps for elements that he predicted were yet to be discovered. He also changed the order for some elements based on atomic weights. Knowledge of isotopes made it possible to explain why the order based on atomic weights was not always correct.

Metals and non-metals

- Metals are elements that react to form positive ions.
- Elements that do not form positive ions are non-metals.

Typical properties of metals and non-metals	
Metals	**Non-metals**
Have high melting / boiling points	Have low melting / boiling points
Conduct heat and electricity	Thermal and electrical insulators
React with oxygen to form alkalis	React with oxygen to form acids
Shiny	Dull
Malleable and ductile	Brittle

SUMMARY

- **Elements in the periodic table are arranged in order of increasing atomic number.**
- **Dmitri Mendeleev left gaps in the periodic table for elements yet to be discovered.**

QUESTIONS

QUICK TEST

1. Potassium has the atomic number 19. What is the electronic structure of potassium?

2. Why did Mendeleev leave gaps in his periodic table?

EXAM PRACTICE

1. Element X is in period 3 and group 2 of the periodic table. Element Y has 14 protons in each atom.

   a) Give the electronic configuration of element X. **[1 mark]**

   b) Will element Y be before or after element X in the periodic table?

   Explain your answer. **[2 marks]**

Groups 0, 1 and 7

Group 0

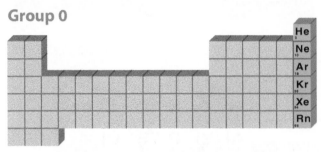

The elements in group 0 are called the **noble gases**. They are chemically inert (unreactive) and do not easily form molecules because their atoms have full outer shells (energy levels) of electrons. The inertness of the noble gases, combined with their low density and non-flammability, means that they can be used in airships, balloons, light bulbs, lasers and advertising signs.

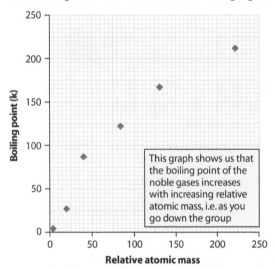

This graph shows us that the boiling point of the noble gases increases with increasing relative atomic mass, i.e. as you go down the group

Group 1 (the alkali metals)

The alkali metals…

- have a low density (lithium, sodium and potassium float on water)

- react with non-metals to form ionic compounds in which the metal ion has a charge of +1

- form compounds that are white solids and dissolve in water to form colourless solutions.

When added to water, the alkali metals float, move about on the surface of the water and effervesce. Sodium melts and potassium reacts with a lilac flame being observed. Hydrogen gas and metal hydroxides are formed. The metal hydroxides dissove in water to form alkaline solutions. For example, for the reaction between sodium and water:

$$2Na_{(s)} + 2H_2O_{(l)} \rightarrow 2NaOH_{(aq)} + H_{2(g)}$$

The alkali metals become more reactive as you go down the group because the outer shell gets further away from the positive attraction of the nucleus. This makes it easier for an atom to lose an electron from its outer shell.

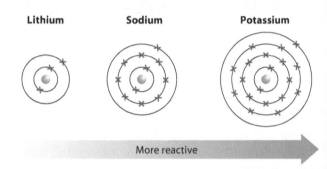

More reactive

Group 7 (the halogens)

The halogens…

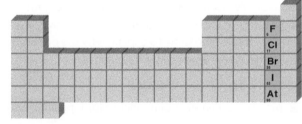

- are non-metals

- consist of diatomic molecules (molecules made up of two atoms)

- react with metals to form ionic compounds where the halide ion has a charge of −1

- form molecular compounds with other non-metals

- form hydrogen halides (e.g. HCl), which dissolve in water, forming acidic solutions.

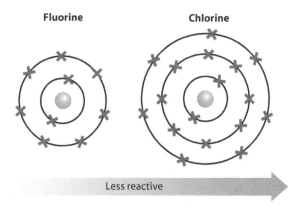

Halogen	State of matter at room temperature	Colour
Fluorine	Gas	Yellow
Chlorine	Gas	Green
Bromine	Liquid	Red / orange
Iodine	Solid	Grey / black

Halogens become less reactive as you go down the group because the outer electron shell gets further away from the attraction of the nucleus, and so an electron is gained less easily.

Fluorine   **Chlorine**

Less reactive

Displacement reactions of halogens

A more reactive halogen will displace a less reactive halogen from an aqueous solution of its metal halide.

For example:

chlorine + potassium bromide → potassium chloride + bromine
Cl_2 + $2KBr$ → $2KCl$ + Br_2

The products of reactions between halogens and aqueous solutions of halide ion salts are as follows.

		Halide salts		
		Potassium chloride, KCl	Potassium bromide, KBr	Potassium iodide, KI
Halogens	Chlorine, Cl_2	No reaction	Potassium chloride + bromine	Potassium chloride + iodine
	Bromine, Br_2	No reaction	No reaction	Potassium bromide + iodine
	Iodine, I_2	No reaction	No reaction	No reaction

SUMMARY

● The elements in group 0 are the noble gases; they are chemically inert.

● Group 1 elements are the alkali metals; they get more reactive as you go down the group.

● Group 7 elements are the halogens; they are non-metals which get less reactive as you go down the group.

QUESTIONS

QUICK TEST

1. Suggest two uses for the noble gases.

2. What are the products of the reaction between potassium and water?

3. What is the trend in reactivity of the halogens as you go down the group?

EXAM PRACTICE

1. A small piece of sodium is added to a trough of water.

   a) What will be observed during this reaction? **[3 marks]**

   b) Write a balanced symbol equation for the reaction occurring.

      Include state symbols. **[2 marks]**

2. Chlorine gas is bubbled through sodium bromide solution.

   a) Explain why the colourless solution turns brown after the chlorine gas has been bubbled through. **[2 marks]**

   b) Explain why this displacement reaction occurs. **[2 marks]**

Transition metals

Comparison with group 1 elements

Typical transition metals are chromium, manganese, iron, cobalt, nickel and copper.

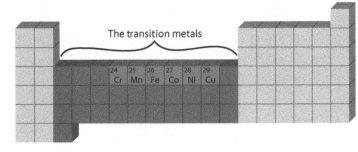

The transition metals

24	25	26	27	28	29
Cr	Mn	Fe	Co	Ni	Cu

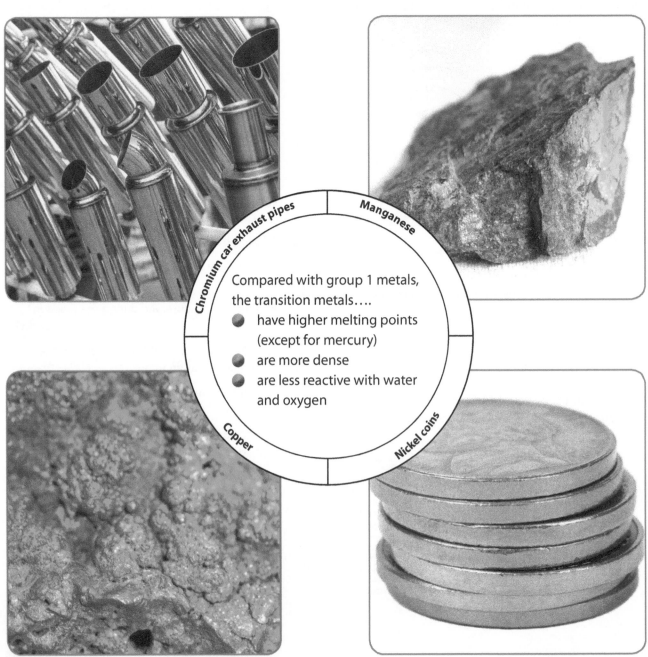

Chromium car exhaust pipes

Manganese

Copper

Nickel coins

Compared with group 1 metals, the transition metals….
- have higher melting points (except for mercury)
- are more dense
- are less reactive with water and oxygen

Properties of transition metals

Many transition elements have ions with different charges. For example, $FeCl_2$ is a compound that contains Fe^{2+} ions and $FeCl_3$ is a compound that contains Fe^{3+} ions.

Compounds of transition metals are also often coloured. For example, potassium manganate(VII), $KMnO_4$, is purple and hydrated copper(II) sulfate, $CuSO_4$, is blue.

The purple colour of potassium manganate(VII) solution

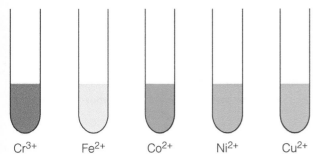

The colours of some other transition metal ions

Cr^{3+} Fe^{2+} Co^{2+} Ni^{2+} Cu^{2+}

Transition metals are also frequently used as catalysts, such as iron in the Haber process (see page 82) and nickel used in the manufacture of margarine.

Margarine

Iron ore

SUMMARY

- The transition metals are located between groups 2 and 3 on the periodic table.
- The transition metals are often used as catalysts, such as iron in the Haber process.

QUESTIONS

QUICK TEST

1. Suggest two properties of transition metals that are different to group 1 metals.

2. Which transition metal compound is purple?

3. Which transition metal is used as a catalyst in the Haber process?

EXAM PRACTICE

1. Sodium is a metal in group 1 in the periodic table. Chromium is a transition metal.

   a) State one physical property that both metals possess. **[1 mark]**

   b) Which one of the two metals:
      i)   forms coloured compounds?
      ii) is less dense than water?
      iii) reacts rapidly with water?
      iv) is frequently used in the production of catalysts?
      v)   forms ions with different charges?
      [5 marks]

Chemical bonding

There are three types of chemical bond:

- ionic
- covalent
- metallic

Ionic bonding

Ionic bonds occur between metals and non-metals. An ionic bond is the electrostatic force of attraction between two oppositely charged ions (called cations and anions).

Ionic bonds are formed when metal atoms transfer electrons to non-metal atoms. This is done so that each atom forms an ion with a full outer shell of electrons.

Example 1

The formation of an ionic bond between sodium and chlorine

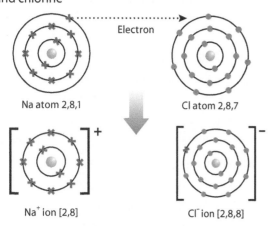

Na atom 2,8,1 Cl atom 2,8,7

Na⁺ ion [2,8]   Cl⁻ ion [2,8,8]

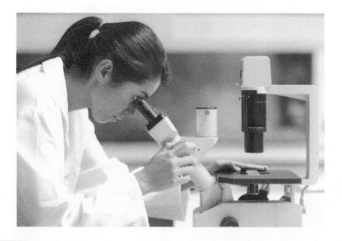

Example 2

The formation of the ionic bond between magnesium and oxygen

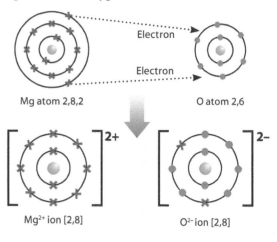

Electron

Electron

Mg atom 2,8,2   O atom 2,6

Mg^{2+} ion [2,8] O^{2-} ion [2,8]

Covalent bonding

Covalent bonds occur between two non-metal atoms. Atoms share electrons so that each atom ends up with a full outer shell of electrons. A covalent bond is a shared pair of electrons, e.g.:

- the formation of a covalent bond between two hydrogen atoms

Hydrogen atoms   **A hydrogen molecule**

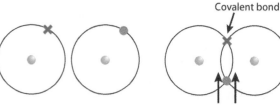

Covalent bond

Outermost shells overlap

- the covalent bonding in methane, CH_4:

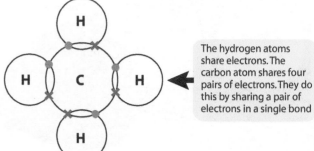

The hydrogen atoms share electrons. The carbon atom shares four pairs of electrons. They do this by sharing a pair of electrons in a single bond

Double covalent bonds occur when two pairs of electrons are shared between atoms, for example in carbon dioxide.

Carbon dioxide

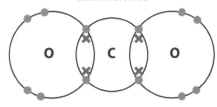

Metallic bonding

Metals consist of giant structures. Each atom loses its outer shell electrons and these electrons become delocalised, i.e. they are free to move through the structure.

The metal cations are arranged in a regular pattern called a lattice.

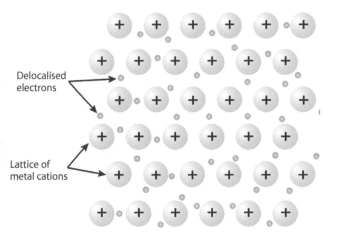

Delocalised electrons

Lattice of metal cations

QUESTIONS

QUICK TEST

1. What type of bonding occurs between non-metal atoms?

2. How many electrons are present in every covalent bond?

3. What is a delocalised electron?

4. What is an ionic bond?

EXAM PRACTICE

1. The electronic configurations of atoms of potassium and sulfur are:

   K 2,8,8,1

   S 2,8,6

   a) Describe the changes in the electronic configurations of potassium and sulfur when these atoms react to form potassium sulfide.　　**[3 marks]**

   b) Give the formula of potassium sulfide.
   　　[1 mark]

2. Ammonia (NH_3) contains covalent bonds.

   a) Draw a dot and cross diagram to show the covalent bonding in ammonia. Show the outer electrons only.　　**[2 marks]**

   b) Explain why neon does not form covalent or ionic bonds.　　**[2 marks]**

3. Thallium is a metal in group 3 of the periodic table.

   Describe the structure of thallium.　　**[2 marks]**

SUMMARY

- There are three types of chemical bond: ionic, covalent and metallic.
- Ionic bonds occur between metals and non-metals.
- Covalent bonds occur between non-metals.

15

Ionic and covalent structures

Structure of ionic compounds

Compounds containing ionic bonds form giant structures. These are held together by strong electrostatic forces of attraction between the oppositely charged ions. These forces act in all directions throughout the lattice.

The diagram below represents a typical giant ionic structure, sodium chloride.

Negatively charged chloride ions

Positively charged sodium ions

The ratio of each ion present in the structure allows the empirical formula of the compound to be worked out. In the diagram above, there are equal numbers of sodium ions and chloride ions. This means that the empirical formula is NaCl.

Sodium chloride

Structure of covalent compounds

Covalently bonded substances may consist of....

- small molecules / simple molecular structures (e.g. Cl_2, H_2O and CH_4)
- large molecules (e.g. polymers – see page 58)
- giant covalent structures (e.g. diamond, graphite and silicon dioxide).

Simple molecular structures

The bonding between hydrogen and carbon in methane can be represented in several ways, as shown here.

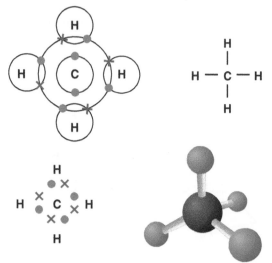

Polymers

Polymers can be represented in the form:

$$\left[\begin{array}{cc} & W \\ | & | \\ C - C \\ | & | \\ Y & X \end{array}\right]_n$$

> **V, W, X and Y represent the atoms bonded to the carbon atoms**

For example, poly(ethene) can be represented as:

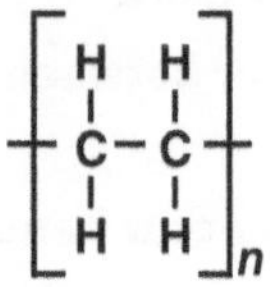

$$\left[\begin{array}{cc} H & H \\ | & | \\ C - C \\ | & | \\ H & H \end{array}\right]_n$$

Giant covalent structures

This is the giant covalent structure of silicon dioxide.

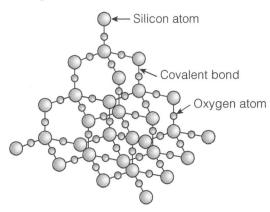

The formula of silicon dioxide is SiO_2 – this can be deduced by looking at the ratio of Si to O atoms in the diagram above.

Models, such as dot-and-cross diagrams, ball-and-stick diagrams and two- / three-dimensional diagrams to represent structures, are limited in value, as they do not accurately represent the structures of materials. For example, a chemical bond is not a solid object as depicted in some models, it is actually an attraction between particles. The relative size of different atoms is often not shown when drawing diagrams.

Diamond – a giant covalent structure

Graphite – a giant covalent structure

SUMMARY

● Ionic compounds are held together by strong electrostatic forces of attraction between oppositely charged ions.

● Substances with covalent bonding can either have simple, molecular, giant covalent or polymer structures.

QUESTIONS

QUICK TEST

1. Name a compound that has a giant ionic structure.

2. What forces hold ionic structures together?

3. What are the three types of covalent substance?

EXAM PRACTICE

1. The diagram below represents one way that the structure of an ionic lattice can be modelled.

   Key

    metal X ion

    chloride (Cl^-) ion

   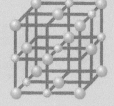

   a) What is the empirical formula of X chloride?

      Explain how you worked this out. **[2 marks]**

   b) Give one way in which the model shown is limited in value. **[1 mark]**

2. Carbon dioxide and silicon dioxide are both oxides of group 4 elements. They contain the same type of chemical bond but they have different structures.

   a) Name the type of bonding present in both carbon dioxide and silicon dioxide. **[1 mark]**

   b) What type of structure does carbon dioxide have? **[1 mark]**

   c) What type of structure does silicon dioxide have? **[1 mark]**

States of matter: properties of compounds

States of matter

The three main states of matter are **solids**, **liquids** and **gases**. Individual atoms do not have the same properties as these bulk substances. The diagram below shows how they can be interconverted and also how the particles in the different states of matter are arranged.

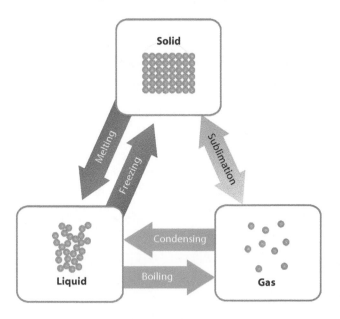

These are physical changes because the particles are either gaining or losing energy and are not undergoing a chemical reaction. Particles in a gas have more energy than in a liquid; particles in a liquid have more energy than in a solid.

Melting and freezing occur at the same temperature. Condensing and boiling also occur at the same temperature. The amount of energy needed to change state depends on the strength of the forces between the particles of the substance.

The stronger the forces between the particles, the higher the melting and boiling points of the substance.

Properties of ionic compounds	
Property	**Explanation**
High melting and boiling points	There are lots of strong bonds throughout an ionic lattice which require lots of energy to break.
Electrical conductivity	Ionic compounds conduct electricity when molten or dissolved in water because the ions are free to move and carry the charge.
	Ionic solids do not conduct electricity because the ions are in a fixed position and are unable to move.

HT The model on the left is limited in value because....

● it does not indicate that there are forces between the spheres

● all particles are represented as spheres

● the spheres are solid.

Properties of small molecules

Substances made up of small molecules are usually gases or liquids at room temperature. They have relatively low melting and boiling points because there are weak (intermolecular) forces that act between the molecules. It is these weak forces and not the strong covalent bonds that are broken when the substance melts or boils.

Substances made up of small molecules do not normally conduct electricity. This is because the molecules do not have an overall electric charge or delocalised electrons.

Polymers

Polymers are very large molecules made up of atoms joined together by strong covalent bonds. The intermolecular forces between polymer molecules are much stronger than in small molecules because the molecules are larger. This is why most polymers are solid at room temperature.

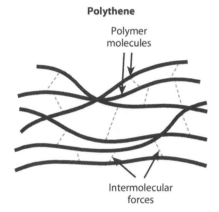

Polythene

Polymer molecules

Intermolecular forces

Giant covalent structures

Substances with a giant covalent structure are solids at room temperature. They have relatively high melting and boiling points. This is because there are lots of strong covalent bonds that need to be broken.

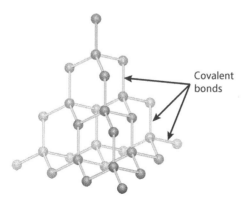

Covalent bonds

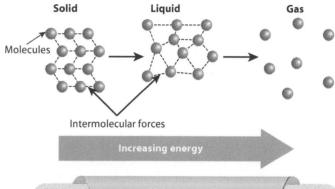

Solid   Liquid   Gas

Molecules

Intermolecular forces

Increasing energy

SUMMARY

- The three main states of matter are solids, liquids and gases.
- Polymers are large molecules made up of atoms joined by strong covalent bonds; most are solids at room temperature.
- Giant covalent structures are solids at room temperature.

QUESTIONS

QUICK TEST

1. What are the three states of matter?

2. What needs to be broken in order to melt a substance made up of small molecules such as water?

EXAM PRACTICE

1. Phosphorus tribromide (PBr_3) has a melting point of −40°C and a boiling point of 175°C.

   a) At what temperature does liquid phosphorus tribromide turn into a solid?   **[1 mark]**

   b) What state of matter is phosphorus tribromide at 100°C?   **[1 mark]**

   c) Name the physical process occurring when the temperature of phosphorus decreases from 200°C to 175°C.   **[1 mark]**

2. Sodium chloride does not conduct electricity at room temperature but does when molten.

   Explain this observation.   **[3 marks]**

Metals, alloys and the structure and bonding of carbon

Structure and properties of metals

Metals have giant structures. Metallic bonding (the attraction between the cations and the delocalised electrons) is strong, meaning that most metals have high melting and boiling points.

The layers are able to slide over each other, which means that metals can be bent and shaped.

Metals are good conductors of electricity because the delocalised electrons are able to move.

The delocalised electrons also transfer energy meaning that they are good thermal conductors.

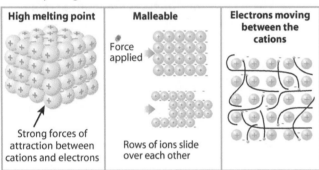

High melting point	Malleable	Electrons moving between the cations

Strong forces of attraction between cations and electrons

Force applied

Rows of ions slide over each other

Alloys

Most metals we use are alloys. Many pure metals (such as gold, iron and aluminium) are too soft for many uses and so are mixed with other materials (usually metals) to make alloys.

The different sizes of atoms in alloys make it difficult for the layers to slide over each other. This is why alloys are harder than pure metals.

Typical alloy structure

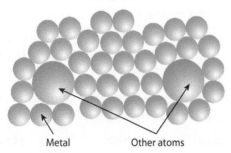

Metal   Other atoms

Structure and bonding of carbon

Carbon has four different structures:

- diamond
- graphite
- graphene
- fullerenes

Diamond

Diamond has a giant covalent (macromolecular) structure where each carbon atom is bonded to four others.

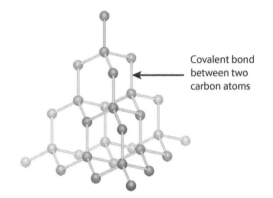

Covalent bond between two carbon atoms

In diamond, there are lots of very strong covalent bonds so diamond…

- is hard
- has a high melting point.

For these reasons, diamond is used in making cutting tools.

There are no free electrons in diamond so it does not conduct electricity.

Graphite

Graphite also has a giant covalent structure, with each carbon atom forming three covalent bonds, resulting in layers of hexagonal rings of carbon atoms. Carbon has four electrons in its outer shell and as only three are used for bonding the other one is delocalised.

The layers in graphite are able to slide over each other because there are only weak intermolecular forces holding them together. This is why graphite is soft and slippery. These properties make graphite suitable for use as a lubricant.

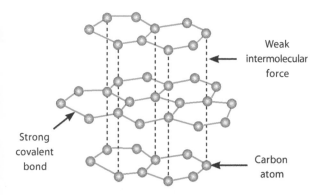

Weak intermolecular force ←

← Strong covalent bond

Carbon atom ←

Like diamond, there are lots of strong covalent bonds in graphite so it has a high melting point.

The delocalised electrons allow graphite to conduct electricity and heat.

Graphene and fullerenes

Graphene is a single layer of graphite and so it is one atom thick.

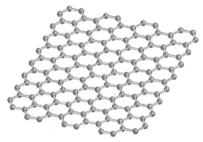

Fullerenes are molecules made up of carbon atoms and they have hollow shapes. The structure of fullerenes is based on hexagonal rings of carbon atoms but they may also contain rings with five or seven carbon atoms.

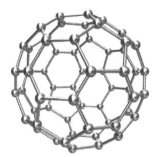

buckminsterfullerene (C_{60}) was the first fullerene to be discovered

Carbon nanotubes are cylindrical fullerenes.

Fullerenes have high…

- tensile strength
- electrical conductivity
- thermal conductivity.

Fullerenes can be used…

- for drug delivery into the body
- as lubricants
- for reinforcing materials, e.g. tennis rackets.

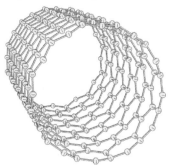

SUMMARY

- **Most metals we use are alloys; alloys are stronger than pure metals.**
- **Diamond and graphite have giant covalent structures; they both have high melting points.**

QUESTIONS

QUICK TEST

1. Why do metals generally have high melting points?

2. Suggest two uses of fullerenes.

EXAM PRACTICE

1. Magnalium is an alloy of aluminium and magnesium. It is used in aircraft manufacture because it is both strong and low in density.

   a) What is an alloy? **[1 mark]**

   b) Explain why magnalium is stronger than pure aluminium. **[2 marks]**

2. Carbon atoms can arrange themselves in different ways. One of those ways is in a form known as graphite.

   a) Describe the structure of graphite. **[3 marks]**

   b) Explain why graphite can conduct electricity. **[2 marks]**

Bulk and surface properties of matter, including nanoparticles

ⓌⓈ Sizes of particles

Nanoscience is the study of structures that are 1–100 nanometres (nm) in size, i.e. a few hundred atoms.

Name of particle	Diameter of particles
Coarse particles / dust (PM_{10})	Between 1×10^{-5} and 2.5×10^{-6} m
Fine particles ($PM_{2.5}$)	Between 1×10^{-7} and 2.5×10^{-6} m (100 and 250 nm)
Nanoparticles	Less than 1×10^{-7} m (100 nm)

PM = particulate matter

As the side of a cube decreases by a factor of 10, the surface area to volume ratio ($\frac{surface\ area}{volume}$) increases by a factor of 10.

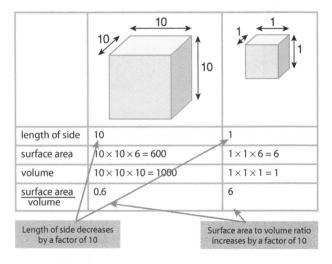

length of side	10	1
surface area	$10 \times 10 \times 6 = 600$	$1 \times 1 \times 6 = 6$
volume	$10 \times 10 \times 10 = 1000$	$1 \times 1 \times 1 = 1$
$\frac{surface\ area}{volume}$	0.6	6

Length of side decreases by a factor of 10

Surface area to volume ratio increases by a factor of 10

Properties of nanoparticles

Nanoparticles may have properties that are different from larger amounts of the same material because of their higher surface area to volume ratio. This may mean that smaller quantities are needed to be effective when compared with materials which have normal-sized particles.

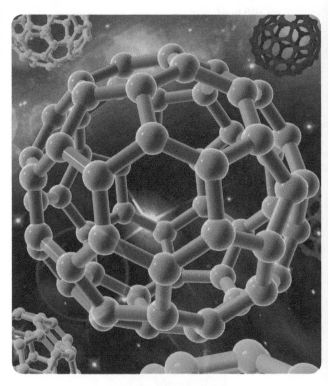

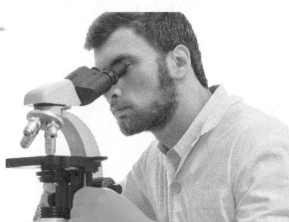

Uses of nanoparticles

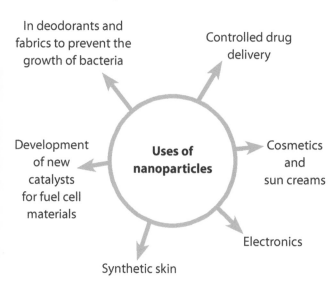

In deodorants and fabrics to prevent the growth of bacteria

Controlled drug delivery

Development of new catalysts for fuel cell materials

Uses of nanoparticles

Cosmetics and sun creams

Electronics

Synthetic skin

Use of nanoparticles in sun creams	
Advantages	**Disadvantages**
Better skin coverage	Potential cell damage in the body
More effective protection from the Sun's ultraviolet rays	Potentially harmful effects on the environment

SUMMARY

- Nanoscience is the study of structures that are 1–100 nanometres in size.
- Nanoparticles have a higher surface area to volume ratio than larger amounts of the same material.
- Nanoparticles have many uses, including in sun creams.

QUESTIONS

QUICK TEST

1. State the size of nanoparticles.

2. What happens to the surface area to volume ratio of a cube when the side of a cube decreases by a factor of 10?

3. Give one advantage of using nanoparticles in sun creams.

EXAM PRACTICE

1. Some manufactures have been adding nanoparticles to socks.

   This is because nanoparticles have antibacterial properties that can reduce the smell of sweaty feet.

   a) State two other uses of nanoparticles.
   [2 marks]

   b) Suggest why some people are cautious about the use of nanoparticles in everyday items such as socks. **[2 marks]**

Mass and equations

Conservation of mass

The total mass of reactants in a chemical reaction is equal to the total mass of the products because atoms are not created or destroyed.

Chemical reactions are represented by balanced symbol equations.

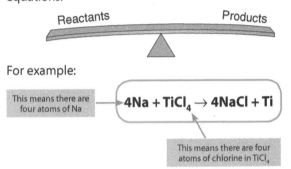

For example:

This means there are four atoms of Na → **4Na + TiCl$_4$ → 4NaCl + Ti**

This means there are four atoms of chlorine in TiCl$_4$

Relative formula mass

The relative formula mass (M$_r$) of a compound is the sum of the relative atomic masses (see the periodic table at the back of this book) of the atoms in the formula.

For example:

- The M$_r$ of MgO is 40   (24 + 16)
- The M$_r$ of H$_2$SO$_4$ is 98   [(2 × 1) + 32 + (4 × 16)]

In a balanced symbol equation, the sum of the relative formula masses of the reactants equals the sum of the relative formula masses of the products.

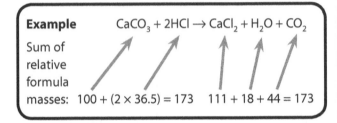

Example	$CaCO_3 + 2HCl \rightarrow CaCl_2 + H_2O + CO_2$
Sum of relative formula masses:	100 + (2 × 36.5) = 173 111 + 18 + 44 = 173

Mass changes when a reactant or product is a gas

During some chemical reactions, there can appear to be a change in mass. When copper is heated its mass actually increases because oxygen is being added to it.

> **copper + oxygen → copper oxide**

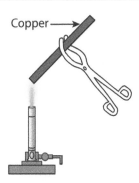

Copper ──→

The mass of copper oxide formed is equal to the starting mass of copper plus the mass of the oxygen that has been added to it.

During a thermal decomposition reaction of a metal carbonate, the final mass of remaining metal oxide solid is less than the starting mass. This is because when the metal carbonate thermally decomposes it releases carbon dioxide gas into the atmosphere.

> **copper carbonate → copper oxide + carbon dioxide**

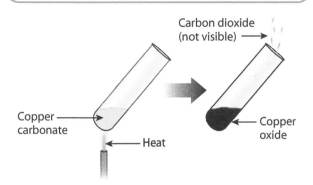

Carbon dioxide (not visible) ──→

Copper carbonate ──→

←── Heat

←── Copper oxide

Example

Starting mass of copper carbonate	= 8.00 g
Final mass of copper oxide	= 5.15 g

Therefore, mass of carbon dioxide
released to the atmosphere   = 2.85 g
(8.00 g − 5.15 g)

SUMMARY

● Balanced equations are used to show chemical reactions.

● Relative formula mass of a compound is the sum of the relative atomic masses of the atoms in the formula.

● Mass changes when a reactant or product is a gas.

QUESTIONS

QUICK TEST

1. In a chemical reaction the total mass of reactants is 9.22 g. Will the expected mass of all the products be lower, higher or the same as 9.22 g?

2. What is relative formula mass?

3. When 6.3 g of zinc carbonate is heated will the mass of remaining solid be lower, higher or the same as 6.3 g?

4. With reference to the periodic table at the back of this book, work out the relative formula mass of the compound $Mg(OH)_2$.

EXAM PRACTICE

1. Magnesium oxide can be made by heating magnesium in air or by heating magnesium carbonate.

   The equation for the reaction that occurs when magnesium carbonate is heated is shown below.

   $$MgCO_{3\,(s)} \rightarrow MgO_{(s)} + CO_{2\,(g)}$$

   a) What is the relative formula mass of magnesium carbonate? **[1 mark]**

   b) If 10 g of magnesium carbonate formed 4.76 g of magnesium oxide upon heating, determine the mass of carbon dioxide produced in this experiment. **[1 mark]**

   c) What name is given to the type of reaction occurring in part b? **[1 mark]**

   d) What will happen to the mass of magnesium metal when it is heated in air? Explain your answer. **[2 marks]**

Moles and masses

Moles and Avogadro's constant

Amounts of chemicals are measured in moles (mol). The number of atoms, molecules or ions in a mole of a given substance is 6.02×10^{23}. This value is known as Avogadro's constant.

Avogadro's constant = 602 000 000 000 000 000 000 000

For example:

- 1 mole of carbon contains 6.02×10^{23} carbon atoms
- 1 mole of sulfur dioxide (SO_2) contains 6.02×10^{23} sulfur dioxide molecules.

Moles and relative formula mass

The mass of one mole of a substance in grams is equal to its relative formula mass (M_r).

For example:

- The mass of one mole of carbon is 12 g.
- The mass of one mole of sulfur dioxide is 64 g.

The number of moles can be calculated using the following formula:

$$M_r = \frac{mass}{moles}$$

For example, the number of moles of carbon in $48\,g = \frac{48}{12} = 4$

By rearranging the above equation, the relative formula mass of a compound can be worked out from the number of moles and mass.

Example

Calculate the relative formula mass of the compound given that 0.23 moles has a mass of 36.8 g.

$$M_r = \frac{mass}{moles} \qquad M_r = \frac{36.8}{0.23} = 160$$

Balanced symbol equations give information about the number of moles of reactants and products. For example:

$$2Mg + O_2 \rightarrow 2MgO$$

This equation tells us that 2 moles of magnesium react with one mole of oxygen to form 2 moles of magnesium oxide.

This means that 48 g of Mg (the mass of 2 moles of Mg) reacts with 32 g of oxygen to form 80 g of magnesium oxide.

	2Mg	+	O_2	→	2MgO
Number of moles reacting	2		1		2
Relative formula mass	24		32		40
Mass reacting/formed (g)	48		32		80

We can use this relationship between mass and moles to calculate reacting masses.

Example: Calculate the mass of magnesium oxide formed when 12 g of magnesium reacts with an excess of oxygen.

	2Mg	+	O_2	→	2MgO
Number of moles reacting	2		1		2
Relative formula mass	24		32		40
Mass reacting/formed (g)	48		32		80
Reacting mass (g)	12				

To get from 48 to 12 we divide by 4

Therefore to find the mass of magnesium oxide formed we divide 80 by 4

The mass of magnesium oxide formed is therefore 20 g.

Empirical formula

The empirical formula is the simplest whole number ratio of each type of atom present in a compound. For example, hexene (C_6H_{12}) has the empirical formula CH_2.

You can work out the empirical formula of a substance from its chemical formula. For example, the empirical formula of ethanoic acid (CH_3COOH) is CH_2O.

The empirical formula of a compound can be calculated from either:

● the percentage composition of the compound by mass

or

● the mass of each element in the compound.

To calculate the empirical formula:

① List all the elements in a compound:

② Divide the data for each element by the relative atomic mass (A_r) of the element (to find the number of moles).

③ Select the smallest answer from step 2 and divide each answer by that result to obtain a ratio.

④ The ratio may need to be scaled up to give whole numbers.

Example 1

What is the empirical formula of a hydrocarbon containing 75% carbon? (Hydrogen = 25%)

	Carbon	:	Hydrogen
①			
②	$\frac{75}{12}$	:	$\frac{25}{1}$
	6.25	:	25
③ ÷ 6.25		:	÷ 6.25
④	1	:	4

So the empirical formula is CH_4.

Example 2

What is the empirical formula of a compound containing 24 g of carbon, 8 g of hydrogen and 32 g of oxygen?

	Carbon	:	Hydrogen	Oxygen
①				
②	$\frac{24}{12}$	:	$\frac{8}{1}$	$\frac{32}{16}$
	2	:	8	2
③ ÷ 2		:	÷ 2	÷ 2
④	1	:	4	1

So the empirical formula is CH_4O.

Molecular formula

The molecular formula is the actual whole number ratio of each type of atom in a compound. It can be the same as the empirical formula or a multiple of the empirical formula. To convert an empirical formula into a molecular formula, you also need to know the relative formula mass of the compound.

Example

A compound has an empirical formula of CH_2 and an M_r of 42. What is its molecular formula?

(A_r for C = 12 and A_r for H = 1)

Work out the relative formula mass of the empirical formula	$= 12 + (2 \times 1) = 14$
Then divide the actual M_r by the empirical formula M_r	$= \frac{42}{14}$
This gives the multiple.	$= 3$

The molecular formula is C_3H_6.

SUMMARY

● Amounts of chemicals are measured in moles (mol).

● The mass of one mole of a substance in grams is equal to its relative formula mass (M_r).

● Empirical formula is the simplest whole number ratio of each type of atom present in a compound.

● Molecular formula is the actual whole number ratio of each type of atom in a compound.

QUESTIONS

QUICK TEST

1. Calculate the empirical formula of a compound containing 0.35 g of lithium and 0.40 g of oxygen.

2. An oxide of phosphorus has the empirical formula P_2O_5 and an M_r of 284. What is its molecular formula?

EXAM PRACTICE

HT 1. The equation for the reaction between aluminium and oxygen is shown below.

$$4Al_{(s)} + 3O_{2(g)} \rightarrow 2Al_2O_{3(s)}$$

Calculate the mass of aluminium oxide formed when 4.32 g of aluminium reacts with oxygen. **[3 marks]**

Percentage yield and atom economy

Percentage yield

Balanced symbol equations allow us to work out the mass of products (yield) we expect to obtain in a chemical reaction. This is known as the 'theoretical yield'. However, the calculated amount of product may not always be obtained, i.e. the percentage yield may not always be 100%.

The actual yield (i.e. the mass of products actually made) compared with the theoretical yield is known as the percentage yield.

$$\% \text{ yield} = \frac{\text{mass of products actually made}}{\text{maximum theoretical mass of product}} \times 100$$

Example

In a chemical reaction the theoretical mass of product is 8.4 g. The mass of product actually obtained was 3.2 g. Calculate the percentage yield for the reaction.

$$\% \text{ yield} = \frac{3.2}{8.4} \times 100 = 38\%$$

HT Example

In the following reaction 12.6 g of nitric acid (HNO_3) reacted with an excess of ammonia (NH_3) to form 12.3 g of ammonium nitrate (NH_4NO_3). Calculate the percentage yield for the reaction.

$$HNO_3 + NH_3 \rightarrow NH_4NO_3$$

Step 1: To calculate the theoretical yield we calculate the expected mass. This is the same calculation as shown on page 26.

	HNO_3	$+ NH_3$	$\rightarrow NH_4NO_3$
Number of moles reacting	1	1	1
M_r	63	17	80
Mass reacting/formed in equation (g)	63 ↘ ÷5		80 ↘ ÷5
Reacting mass (g)	12.6 ↙		16 ↙

The theoretical / expected mass of ammonium nitrate is 16 g.

Step 2: Calculate the percentage yield.

$$\% \text{ yield} = \frac{12.3}{16} \times 100$$
$$= 76.9\% \text{ (to 1 d.p.)}$$

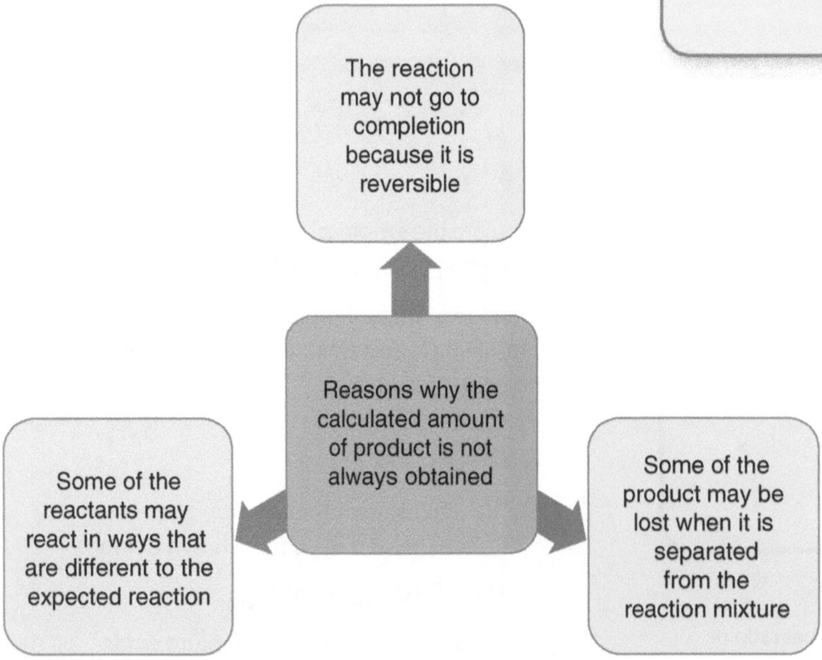

The reaction may not go to completion because it is reversible

Reasons why the calculated amount of product is not always obtained

Some of the reactants may react in ways that are different to the expected reaction

Some of the product may be lost when it is separated from the reaction mixture

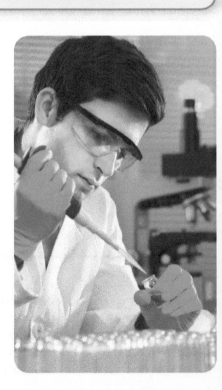

Atom economy

Atom economy is a measure of the amount of starting materials that end up as useful products, i.e. a measure of how much starting material is not wasted.

Reactions with a high atom economy are important for sustainable development and are more economic because of the lack of waste formed and the lower demand for raw materials.

The percentage atom economy of a reaction is calculated from the balanced equation of the reaction and using the following equation:

$$\frac{\text{relative formula mass of desired product from equation}}{\text{sum of relative formula masses of all reactants from equation}} \times 100$$

Example 1

Calculate the atom economy for the production of hydrogen from the following reaction.

$$Mg + 2HCl \rightarrow MgCl_2 + H_2$$

In this case, the 'desired product' is hydrogen. Hydrogen has a relative formula mass (M_r) of 2.

The sum of the relative formula masses of the reactants is $24 + (2 \times 36.5) = 97$.

$$\text{Atom economy} = \frac{2}{97} \times 100$$
$$= 2.1\% \text{ (to 1 d.p.)}$$

Example 2

Ethene (C_2H_4) can be made by the cracking of decane ($C_{10}H_{22}$) according to the equation:

$$C_{10}H_{22} \rightarrow 2C_2H_4 + C_6H_{14}$$

Calculate the atom economy for this method of producing ethene.

M_r ethene = 28

M_r decane ($C_{10}H_{22}$) = 142

$$\text{Atom economy} = \frac{2 \times 28}{142} \times 100$$

We need to multiply by 2 because there is a 2 in front of C_2H_4 in the balanced equation for the reaction

$$\text{Atom economy} = 39.4\% \text{ (to 1 d.p.)}$$

- Percentage yield is the ratio of mass of product obtained to mass of product expected. Balanced symbol equations allow us to work this out.
- Atom economy is a measure of the amount of starting materials that end up as useful products.

QUESTIONS

QUICK TEST

1. What is meant by atom economy?

2. Suggest why the percentage yield for a reaction may not be 100%.

EXAM PRACTICE

1. In the Ostwald process, nitric acid, HNO_3, can be produced by reacting nitrogen dioxide with water. The equation for this reaction is:

$$3NO_{2\,(g)} + H_2O_{\,(l)} \rightarrow 2HNO_{3\,(aq)} + NO_{\,(g)}$$

   a) Calculate the atom economy for this reaction. **[2 marks]**

   b) Calculate the percentage yield for this reaction if 12.1 g of nitric acid is obtained and the expected amount was 17.5 g.

   Give your answer to three significant figures. **[2 marks]**

   c) Nitrogen dioxide can also be converted to nitric acid by the following reaction:

$$4NO_{2\,(g)} + O_{2\,(g)} + 2H_2O_{\,(l)} \rightarrow 4HNO_{3\,(aq)}$$

   With reference to the atom economy, state and explain the environmental advantage that this method offers compared to the Ostwald process. **[3 marks]**

Moles, solutions and masses

Concentration of solutions in g/dm³

Many chemical reactions take place in solutions. The concentration of a solution can be measured in mass of solute per given volume of solution, e.g. grams per dm³ (1 dm³ = 1000 cm³).

For example, a solution of 5 g/dm³ has 5 g of solute dissolved in 1 dm³ of water. It has half the concentration of a 10 g/dm³ solution of the same solute.

The mass of solute in a solution can be calculated if the concentration and volume of solution are known.

Example

Calculate the mass of solute in 250 cm³ of a solution whose concentration is 8 g/dm³.

Step 1: Divide the mass by 1000 (this gives you the mass of solute in 1 cm³).

$8 \div 1000 = 0.008$ g/cm³

Step 2: Multiply this value by the volume specified.

$0.008 \times 250 = 2$ g

Concentrations of solutions in mol/dm³

Solution concentrations can also be measured in mol/dm³ (i.e. how many moles of a solute are dissolved in 1 dm³ (1000 cm³) of water).

$$moles = \frac{mass}{relative\ formula\ mass}$$

If we know the concentration of a solution in g/dm³, we can work out the concentration in mol/dm³.

Example

What is the concentration in mol/dm³ of a 5 g/dm³ solution of NaOH?

The relative formula mass of NaOH is 40.

The number of moles of NaOH present in 5 g is $\frac{5}{40} = 0.125$.

Therefore the concentration is 0.125 mol/dm³.

WS Acid–alkali titrations

A titration is an accurate technique that you can use to find out how much of an acid is needed to neutralise an alkali of known concentration (called a standard solution).

When neutralisation takes place, the hydrogen ions (H⁺) from the acid join with the hydroxide ions (OH⁻) from the alkali to form water (neutral pH).

hydrogen ion + hydroxide ion → water molecule
$H^+_{(aq)}$ + $OH^-_{(aq)}$ → $H_2O_{(l)}$

Use the following titration method:

1. Wash and rinse a pipette with the alkali that you will use.

2. Use the pipette to measure out a known and accurate volume of the alkali.

3. Place the alkali in a clean, dry conical flask. Add a few drops of a suitable indicator, e.g. phenolphthalein.

4. Place the acid in a burette that has been carefully washed and rinsed with the acid. Take a reading of the volume of acid in the burette (initial reading). Ensure the jet space is filled with acid.

5. Carefully add the acid to the alkali (while swirling the flask) until the indicator changes colour to show neutrality. This is called the end point. Take a reading of the volume of acid in the burette (final reading).

6. Calculate the volume of acid added (i.e. subtract the initial reading from the final reading).

This method can be repeated to check results.

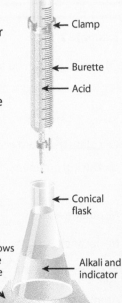

Clamp

Burette

Acid

Conical flask

White tile (allows you to see the colour change clearly)

Alkali and indicator

Titrations can also be used to find the **concentration** of an **acid** or **alkali** providing you know either…

● the relative **volumes** of acid and alkali used

or

● the **concentration** of the other acid or alkali.

It will help if you break down the calculation.

1 Write down a **balanced equation** for the reaction to determine the ratio of moles of acid to alkali involved.

2 Calculate the number of moles in the solution of known volume and concentration. (You will know the number of moles in the other solution from your previous calculation.)

3 Calculate the concentration of the other solution using this formula:

$$\text{concentration of solution (mol/dm}^3 \text{ or M)} = \frac{\text{number of moles of solute (mol)}}{\text{volume of solution (dm}^3)}$$

Example

A titration is carried out and 0.04 dm³ hydrochloric acid neutralises 0.08 dm³ sodium hydroxide of concentration 1 mol dm⁻³. Calculate the concentration of the hydrochloric acid.

> Write the balanced symbol equation for the reaction

$$HCl_{(aq)} + NaOH_{(aq)} \rightarrow NaCl_{(aq)} + H_2O_{(l)}$$

> You can see that 1 mol of HCl neutralises 1 mol of NaOH

Rearrange the formula:

number of moles of NaOH = concentration of NaOH × volume of NaOH

$$= 1 \text{ mol/dm}^3 \times 0.08 \text{ dm}^3$$

$$= 0.08 \text{ mol}$$

> Number of moles of HCl used up in the reaction is also 0.08 mol

Now calculate the concentration of HCl:

$$\text{concentration of HCl} = \frac{\text{number of moles of HCl}}{\text{volume of HCl}}$$

$$= \frac{0.08 \text{ mol}}{0.04 \text{ dm}^3}$$

$$= 2 \text{ mol/dm}^3$$

SUMMARY

● Concentration of a solution can be measured in mass of solute per given volume of solution.

● A titration is a technique to find out how much acid is needed to neutralise a standard solution.

QUESTIONS

QUICK TEST

1. Calculate the mass of solute in 140 cm³ of a solution whose concentration is 6 g/dm³.

2. What is the concentration in mol/dm³ of a 19.6 g/dm³ solution of H_2SO_4?

EXAM PRACTICE

1. A student found a bottle of sulfuric acid, H_2SO_4, that had on its label: 'concentration = 5 g/dm³'

   In order to test the accuracy of the concentration of the sulfuric acid the student performed a titration. 25.0 cm³ portions of the sulfuric acid were transferred to a conical flask. A few drops of indicator were added. A burette was filled with sodium hydroxide of concentration 0.10 mol/dm³. The sulfuric acid was titrated with the sodium hydroxide until the indicator changed colour. The experiment was repeated. The average burette reading was 18.60 cm³.

   The equation for the reaction between sulfuric acid and sodium hydroxide is:

   $$H_2SO_{4\,(aq)} + 2NaOH_{(aq)} \rightarrow Na_2SO_{4\,(aq)} + 2H_2O_{(l)}$$

   a) Calculate the number of moles of sodium hydroxide used during each titration.
   [2 marks]

   b) Deduce the number of moles of sulfuric acid reacting during each titration. **[2 marks]**

   c) Calculate the concentration of the sulfuric acid in mol/dm³. **[2 marks]**

   d) Comment on the accuracy of the label on the bottle of the sulfuric acid. **[3 marks]**

HT Moles and gases

Volumes of gases

The volume of one mole of any gas measured at room temperature and pressure (20°C and 1 atmosphere pressure) is 24 dm³. Based on this fact, equal volumes of gases contain the same number of moles (when compared under the same temperature and pressure).

number of moles of gas = $\dfrac{\text{volume}}{24}$

measured in dm³

Example

At 20°C and 1 atmosphere pressure, how many moles are present in 18 dm³ of carbon dioxide gas?

number of moles of gas $= \dfrac{18}{24}$

$= 0.75$

Gas volumes and equations

The volumes of gaseous reactants and products can be calculated from the balanced equation for the reaction.

Example

Calculate the volume of hydrogen gas formed (at room temperature and pressure) when 3 g of magnesium reacts with excess hydrochloric acid. The equation for the reaction is:

$$Mg_{(s)} + 2HCl_{(aq)} \rightarrow MgCl_{2(aq)} + H_{2(g)}$$

Step 1: Work out the mass of hydrogen formed.

The calculations on page 28 show how to work out the mass of hydrogen formed.

Number of moles reacting	1	2	1	1
Relative formula mass	24	36.5	95	2
Mass reacting / formed in equation (g)	24 ÷8	73	95	2 ÷8
Reacting mass (g)	3			0.25

Step 2: Work out the number of moles of hydrogen formed.

moles $= \dfrac{\text{mass}}{M_r}$

Therefore the number of moles of hydrogen formed $= \dfrac{0.25}{2} = 0.125$

Step 3: Work out the volume of hydrogen formed, by rearranging:

number of moles of gas $= \dfrac{\text{volume}}{24}$

we can work out the volume of hydrogen formed.

number of moles × 24 = volume

$0.125 \times 24 = 3$ dm³

Therefore, the volume of hydrogen formed in this experiment is 3 dm³ (3000 cm³).

Using moles to balance equations

The masses of reactants / products in an equation and the M_r values can be used to work out the balancing numbers in a symbol equation.

Example

Balance the equation below given that 8 g of CH_4 reacts with 32 g of oxygen to form 22 g of CO_2 and 18 g of H_2O

$$...CH_4 + ...O_2 \rightarrow ...CO_2 + ...H_2O$$

Chemical	CH_4	O_2	CO_2	H_2O
Mass (from question)	8	32	22	18
M_r	16	32	44	18
Moles = $\dfrac{mass}{M_r}$	$\dfrac{8}{16} = 0.5$	$\dfrac{32}{32} = 1$	$\dfrac{22}{44} = 0.5$	$\dfrac{18}{18} = 1$

We can make this a whole number ratio by dividing all answers by the smallest answer

÷ 0.5				
	$\dfrac{0.5}{0.5} = 1$	$\dfrac{1}{0.5} = 2$	$\dfrac{0.5}{0.5} = 1$	$\dfrac{1}{0.5} = 2$

The balanced equation is therefore:

$$......CH_4 + ..2..O_2 \rightarrow ......CO_2 + ..2..H_2O$$

QUESTIONS

QUICK TEST

1. At 20 °C and 1 atmosphere pressure, how many moles are present in 15.3 dm³ of neon gas?

2. Balance the equation below for the reaction that occurs when 7 g of silicon reacts with 35.5 g of chlorine to form 42.5 g of silicon chloride.

   $$..........Si + ...Cl_2 \rightarrow ....SiCl_4$$

EXAM PRACTICE

1. Calculate the minimum volume, in dm³, of oxygen required to completely combust 3 dm³ of methane gas.　　**[3 marks]**

   $$CH_{4\,(g)} + 2O_{2\,(g)} \rightarrow CO_{2\,(g)} + 2H_2O_{\,(g)}$$

2. Lithium reacts with nitrogen gas as shown in the equation below.

   $$6Li_{\,(s)} + N_{2\,(g)} \rightarrow 2Li_3N_{\,(s)}$$

   Calculate the minimum volume of nitrogen, in cm³ required to react with 4.2g of lithium.　　**[3 marks]**

Reactivity of metals and metal extraction

Reaction of metals with oxygen

Many metals react with oxygen to form metal oxides, for example:

> **copper + oxygen → copper oxide**

These reactions are called oxidation reactions. Oxidation reactions take place when a chemical gains oxygen.

When a substance loses oxygen it is called a reduction reaction.

The reactivity series

When metals react they form positive ions. The more easily the metal forms a positive ion the more reactive the metal.

Calcium and magnesium are both in group 2 of the periodic table so will form 2+ ions when they react. Calcium is more reactive than magnesium so it has a greater tendency/is more likely to form the 2+ ion.

Decreasing reactivity →

Metal	Reaction with water	Reaction with acid
Potassium	very vigorous	explosive
Sodium	vigorous	dangerous
lithium	steady	very vigorous
calcium	steady fizzing and bubbling	vigorous
magnesium	slow reaction	steady fizzing and bubbling
aluminium	slow reaction	steady fizzing and bubbling
*Carbon		
zinc	very slow reaction	gentle fizzing and bubbling
iron	extremely slow	slight fizzing and bubbling
*Hydrogen		
copper	no reaction	no reaction
silver	no reaction	no reaction
gold	no reaction	no reaction

* included for comparison (because these non-metals can displace metals)

The reactivity series can also be used to predict displacement reactions.

> **zinc + copper oxide → zinc oxide + copper**

In this reaction…

- zinc displaces (i.e. takes the place of) copper
- zinc is oxidised (i.e. it gains oxygen)
- copper oxide is reduced (i.e. it loses oxygen).

HT Oxidation and reduction in terms of electrons

Oxidation and reduction can also be defined in terms of electrons:

O oxidation
I is
L loss (of electrons)
R reduction
I is
G gain (of electrons)

This mnemonic can be useful to work out what is being oxidised and reduced in displacement reactions, for example:

magnesium +	copper sulfate	→	magnesium sulfate	+ copper
Mg +	$CuSO_4$	→	$MgSO_4$	+ Cu

The ionic equation for this reaction is:

Mg **loses electrons** to become Mg^{2+}

Oxidation

$$Mg + Cu^{2+} \rightarrow Mg^{2+} + Cu$$

Reduction

Cu^{2+} **gains electrons** to become Cu

Extraction of metals and reduction

Unreactive metals, such as gold, are found in the Earth's crust as pure metals. Most metals are found as compounds and chemical reactions are required to extract the metal. The method of extraction depends on the position of the metal in the reactivity series.

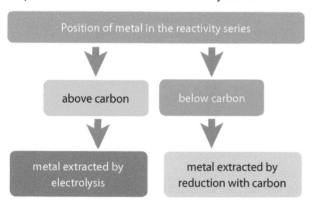

Position of metal in the reactivity series

above carbon → metal extracted by electrolysis

below carbon → metal extracted by reduction with carbon

For example, iron is found in the earth as iron(III) oxide, Fe_2O_3. The iron(III) oxide can be reduced by reacting it with carbon.

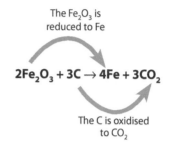

The Fe_2O_3 is reduced to Fe

$$2Fe_2O_3 + 3C \rightarrow 4Fe + 3CO_2$$

The C is oxidised to CO_2

SUMMARY

- An oxidation reaction is when a metal reacts with oxygen. A reduction reaction is when a substance loses oxygen.

- The reactivity series can be used to predict displacement reactions.

QUESTIONS

QUICK TEST

1. Suggest two metals from the reactivity series that are extracted by reduction with carbon.

2. Which metal is the most reactive – magnesium or iron?

3. Write a word equation for the oxidation of aluminium to form aluminium oxide.

4. Both sodium and lithium react to form 1+ ions. Which one of these metals is more likely to form this ion?

EXAM PRACTICE

1. Magnesium reacts with copper(II) oxide to form magnesium oxide and copper.

   a) Explain why this reaction takes place. **[1 mark]**

   b) In this reaction which substance has been reduced?

   Explain your answer. **[2 marks]**

Reactions of acids

Reactions of acids with metals

Acids react with metals that are above hydrogen in the reactivity series to make salts and hydrogen, for example:

> magnesium + hydrochloric acid → magnesium chloride + hydrogen

This is a salt

HT The reactions of metals with acids are redox reactions. The ionic equation for the reaction of magnesium with hydrochloric acid is:

$$Mg + 2H^+ \rightarrow Mg^{2+} + H_2$$

The metal (in this case magnesium) is oxidised, i.e. it loses electrons.

The hydrogen ions are reduced, i.e. they gain electrons.

Mg is **oxidised**, i.e. it loses electrons (to form Mg^{2+})

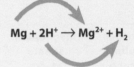

$$Mg + 2H^+ \rightarrow Mg^{2+} + H_2$$

H^+ is **reduced**, i.e. it gains electrons (to form H_2)

Neutralisation of acids and the preparation of salts

Acids can be neutralised by the following reactions.

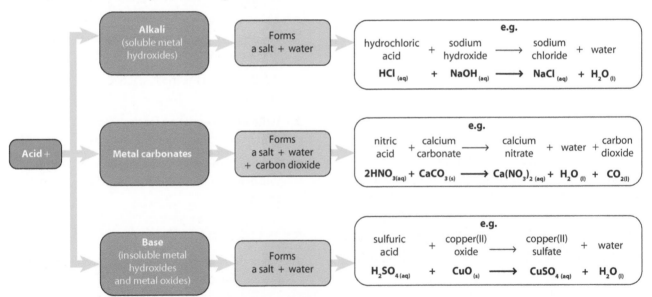

Acid +

Alkali (soluble metal hydroxides) → Forms a salt + water → **e.g.**

hydrochloric acid + sodium hydroxide ⟶ sodium chloride + water

$$HCl_{(aq)} + NaOH_{(aq)} \longrightarrow NaCl_{(aq)} + H_2O_{(l)}$$

Metal carbonates → Forms a salt + water + carbon dioxide → **e.g.**

nitric acid + calcium carbonate ⟶ calcium nitrate + water + carbon dioxide

$$2HNO_{3(aq)} + CaCO_{3(s)} \longrightarrow Ca(NO_3)_{2(aq)} + H_2O_{(l)} + CO_{2(l)}$$

Base (insoluble metal hydroxides and metal oxides) → Forms a salt + water → **e.g.**

sulfuric acid + copper(II) oxide ⟶ copper(II) sulfate + water

$$H_2SO_{4(aq)} + CuO_{(s)} \longrightarrow CuSO_{4(aq)} + H_2O_{(l)}$$

The first part of the salt formed contains the positive ion (usually the metal) from the alkali, base or carbonate followed by…

- chloride if hydrochloric acid was used
- sulfate if sulfuric acid was used
- nitrate if nitric acid was used.

For example, when calcium hydroxide is reacted with sulfuric acid, the salt formed is calcium sulfate.

Making salts

Salts can be either soluble or insoluble. The majority of salts are soluble.

The general rules for deciding whether a salt will be soluble are as follows.

- All common sodium, potassium and ammonium salts are soluble.
- All nitrates are soluble.
- All common chlorides, except silver chloride, are soluble.
- All common sulfates, except barium and calcium, are soluble.
- All common carbonates are insoluble, except potassium, sodium and ammonium.

(WS) Preparation of soluble salts

Soluble salts can be prepared by the following method.

For example, copper(II) sulfate crystals can be made by reacting copper(II) oxide with sulfuric acid.

Add copper(II) oxide to sulfuric acid and stir

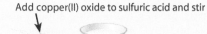

Sulfuric acid

Copper(II) sulfate

(caption continued) Copper(II) sulfate

SUMMARY

- Acids can be neutralised by reacting the acid with an alkali, a metal carbonate or a base.
- Salts can be either soluble or insoluble. Most salts are soluble.
- Soluble salts can be made by reacting acid with a solid, filtering off the excess solid and then obtaining the salt by crystallisation.

QUESTIONS

QUICK TEST

1. What is crystallisation?

HT 2. Identify the species that is oxidised in the following reaction.

$$Fe + 2H^+ \rightarrow Fe^{2+} + H_2$$

3. Write a word equation for the reaction that occurs when aluminium reacts with sulfuric acid.

4. Name the salt formed when zinc oxide reacts with nitric acid.

5. Which one of the following salts is insoluble?
   sodium carbonate
   calcium sulfate
   copper(II) nitrate

EXAM PRACTICE

1. Magnesium oxide powder reacts with hydrochloric acid to form water and the soluble salt magnesium chloride.

   a) Write a balanced symbol equation for this reaction.

   Include state symbols. **[2 marks]**

   b) Describe how a pure, dry sample of magnesium chloride can be prepared from magnesium oxide and hydrochloric acid.

   Include any necessary equipment in your answer. **[5 marks]**

pH, neutralisation, acid strength and electrolysis

Indicators, the pH scale and neutralisation reactions

Indicators are useful dyes that become different colours in acids and alkalis.

Indicator	Colour in acid	Colour in alkali
Litmus	Red	Blue
Phenolphthalein	Colourless	Pink
Methyl orange	Pink	Yellow

The pH scale measures the acidity or alkalinity of a solution. The pH scale runs from 0 to 14 and the pH of a solution can be measured using universal indicator or a pH probe.

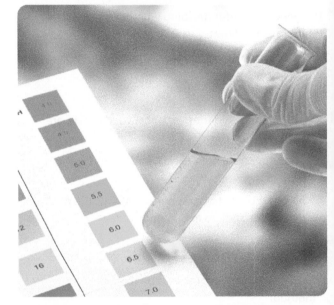

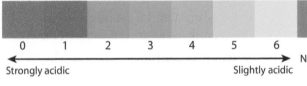

| 0 | 1 | 2 | 3 | 4 | 5 | 6 | 7 | 8 | 9 | 10 | 11 | 12 | 13 | 14 |

Strongly acidic ← Slightly acidic → Neutral ← Slightly alkaline → Strongly alkaline

Acids are solutions that contain hydrogen ions (H^+). The higher the concentration of hydrogen ions, the more acidic the solution (i.e. the lower the pH).

Alkalis are solutions that contain hydroxide ions (OH^-). The higher the concentration of hydroxide ions, the more alkaline the solution (i.e. the higher the pH).

When an acid is neutralised by an alkali, the hydrogen ions from the acid react with the hydroxide ions in the alkali to form water.

$$H^+_{(aq)} + OH^-_{(aq)} \rightarrow H_2O_{(l)}$$

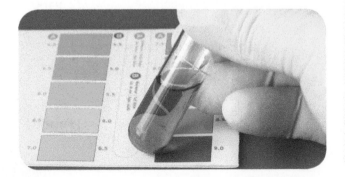

HT Strong acids, such as hydrochloric, sulfuric and nitric acids, are those that completely ionise in aqueous solution. For example:

$$HCl_{(aq)} \rightarrow H^+_{(aq)} + Cl^-_{(aq)}$$

This means dissolved in water

Weak acids, such as ethanoic, citric and carbonic acids, only partially ionise in water. For example:

$$CH_3COOH_{(aq)} \rightleftharpoons CH_3COO^-_{(aq)} + H^+_{(aq)}$$

This sign means that the reaction is reversible, i.e. that the acid does not fully ionise

The pH of an acid is a measure of the concentration of hydrogen ions. When two different acids of the same concentration have different pH values the strongest acid will have the lowest pH.

The pH scale is a logarithmic scale. As the pH decreases by 1 unit (e.g. from 3 to 2) the hydrogen ion concentration increases by a factor of 10.

Electrolysis

An electric current is the flow of electrons through a conductor but it can also flow by the movement of ions through a solution or a liquid.

Covalent compounds do not contain free electrons or ions that can move. So they will not conduct electricity when solid, liquid, gas or in solution.

The ions in:

● an **ionic solid** are fixed and cannot move

● an **ionic substance** that is **molten** or in **solution** are free to move.

Electrolysis is a chemical reaction that involves passing electricity through an electrolyte. An electrolyte is a liquid that conducts electricity. Electrolytes are either molten ionic compounds or solutions of ionic compounds. Electrolytes are decomposed during electrolysis.

● The positive ions (**cations**) move to, and discharge at, the negative electrode (**cathode**).

● The negative ions (**anions**) move to, and discharge at, the positive electrode (**anode**).

Electrons are removed from the anions at the anode. These electrons then flow around the circuit to the cathode and are transferred to the cations.

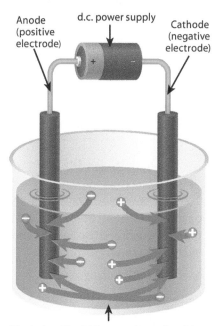

Electrolyte (liquid that conducts electricity and decomposes in electrolysis)

SUMMARY

● The pH scale measures the acidity or alkalinity of a solution.

● Acids are solutions that contain hydrogen ions; alkalis are solutions that contain hydroxide ions.

● Electrolysis is a chemical reaction that involves passing electricity through an electrolyte.

QUESTIONS

QUICK TEST

1. What is an electrolyte?

2. Which ion is responsible for solutions being acidic?

3. What type of substance has a pH of more than 7?

HT 4. What is the difference between a strong acid and a weak acid?

5. What name is given to positive ions?

EXAM PRACTICE

1. Sodium hydroxide is a common laboratory alkali. In an experiment, hydrochloric acid was added to sodium hydroxide solution containing phenolphthalein indicator until the indicator changed colour.

   a) What colour will the phenolphthalein be at the beginning and end of the experiment? **[2 marks]**

   b) Name the ion that causes solutions to be alkaline. **[1 mark]**

   c) Write the ionic equation for the reaction between sodium hydroxide and hydrochloric acid. **[1 mark]**

   HT d) Hydrochloric acid is a strong acid.

   Explain the meaning of the term 'strong acid'. **[2 marks]**

Applications of electrolysis

Electrolysis of molten ionic compounds

When an ionic compound melts, electrostatic forces between the charged ions in the crystal lattice are broken down, meaning that the ions are free to move.

When a direct current is passed through a molten ionic compound:

- positively charged ions are attracted towards the **negative electrode** (cathode)

- negatively charged ions are attracted towards the **positive electrode** (anode).

For example, in the electrolysis of molten lead bromide:

- positively charged lead ions are attracted towards the cathode, forming lead

- negatively charged bromide ions are attracted towards the anode, forming bromine.

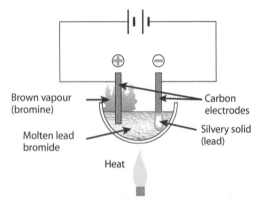

Brown vapour (bromine)
Carbon electrodes
Molten lead bromide
Silvery solid (lead)
Heat

When ions get to the oppositely charged electrode they are **discharged**, i.e. they lose their charge. For example, in the electrolysis of molten lead bromide, the non-metal ion loses electrons to the positive electrode to form a bromine atom. The bromine atom then bonds with a second atom to form a bromine molecule.

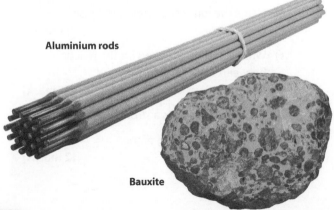

Aluminium rods

Bauxite

(ws) Using electrolysis to extract metals

Aluminium is the most abundant metal in the Earth's crust. It must be obtained from its ore by electrolysis because it is too reactive to be extracted by heating with carbon. The electrodes are made of graphite (a type of carbon). The aluminium ore (bauxite) is purified to leave aluminium oxide, which is then melted so that the ions can move. Cryolite is added to increase the conductivity and lower the melting point.

When a current passes through the molten mixture:

- positively charged aluminium ions move towards the negative electrode (**cathode**) and form aluminium

- negatively charged oxygen ions move towards the positive electrode (**anode**) and form oxygen.

The positive electrodes gradually wear away (because the graphite electrodes react with the oxygen to form carbon dioxide gas). This means they have to be replaced every so often. Extracting aluminium can be quite an expensive process because of the cost of the large amounts of electrical energy needed to carry it out.

Electrolysis of aqueous solutions

When a solution undergoes electrolysis, there is also water present. During electrolysis water molecules break down into hydrogen ions and hydroxide ions.

$$H_2O_{(l)} \rightarrow H^+_{(aq)} + OH^-_{(aq)}$$

This means that when an aqueous compound is electrolysed there are two cations present (H^+ from water and the metal cation from the compound) and two anions present (OH^- from water and the anion from the compound).

For example, in copper(II) sulfate solution…

- cations present: Cu^{2+} and H^+
- anions present: SO_4^{2-} and OH^-

At the positive electrode (anode): *Oxygen is produced unless the solution contains halide ions*. In this case the oxygen is produced.

At the negative electrode (cathode): *The least reactive element is formed*. The reactivity series on page 34 will be helpful here. In this case, hydrogen is formed.

Example

What are the three products of the electrolysis of sodium chloride solution?

Cations present: Na^+ and H^+

Q. What happens at the cathode?

A. Hydrogen is less reactive than sodium therefore hydrogen gas is formed.

Anions present: Cl^- and OH^-

Q. What happens at the anode?

A. A halide ion (Cl^-) is present therefore chlorine will be formed.

The sodium ions and hydroxide ions stay in solution (i.e. sodium hydroxide solution remains).

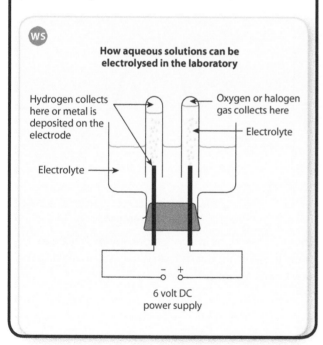

WS

How aqueous solutions can be electrolysed in the laboratory

Hydrogen collects here or metal is deposited on the electrode

Oxygen or halogen gas collects here

Electrolyte

Electrolyte

6 volt DC power supply

Solution	Product at cathode	Product at anode
copper chloride	copper	chlorine
sodium sulfate	hydrogen	oxygen
water (diluted with sulfuric acid to aid conductivity)	hydrogen	oxygen

---

HT Half-equations

During electrolysis, the cation that is discharged at the cathode gains electrons (is reduced) to form the element. For example:

$$Cu^{2+} + 2e^- \rightarrow Cu$$

$$2H^+ + 2e^- \rightarrow H_2$$

At the anode the anion loses electrons (is oxidised). For example:

$$2Cl^- \rightarrow Cl_2 + 2e^-$$

this can also be written as $2Cl^- - 2e^- \rightarrow Cl_2$

$$4OH^- \rightarrow O_2 + 2H_2O + 4e^-$$

or $4OH^- - 4e^- \rightarrow O_2 + 2H_2O$

SUMMARY

- When a direct current is passed through a molten ionic compound, positively charged ions move towards the cathode, and negatively charged ions move towards the anode.
- Electrolysis can be used to obtain aluminium from its ore.
- During electrolysis, water molecules break down into hydrogen ions and hydroxide ions.

QUESTIONS

QUICK TEST

1. When molten copper chloride is electrolysed, what will be formed at the cathode and anode?

2. What ions do water molecules break down into during electrolysis?

EXAM PRACTICE

1. Name the products and state at which electrode they will be formed when copper sulfate solution is electrolysed. **[2 marks]**

Energy changes in reactions

Exothermic and endothermic reactions

Energy is not created or destroyed during chemical reactions, i.e. the amount of energy in the universe at the end of a chemical reaction is the same as before the reaction takes place.

Type of reaction	Is energy given out or taken in?	What happens to the temperature of the surroundings?
Exothermic	out	increases
Endothermic	in	decreases

Examples of exothermic reactions include…

- combustion
- neutralisation
- many oxidation reactions
- precipitation reactions
- displacement reactions.

Everyday applications of exothermic reactions include self-heating cans and hand warmers.

Examples of endothermic reactions include…

- thermal decomposition
- the reaction between citric acid and sodium hydrogen carbonate.

Some changes, such as dissolving salts in water, can be either exothermic or endothermic. Some sports injury packs are based on endothermic reactions.

Reaction profiles

For a chemical reaction to occur, the reacting particles must collide together with sufficient energy. The minimum amount of energy that the particles must have in order to react is known as the 'activation energy'.

Reaction profiles can be used to show the relative energies of reactants and products, the activation energy and the overall energy change of a reaction.

Reaction profile for an exothermic reaction

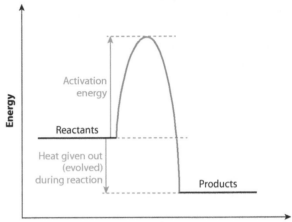

Chemical reactions in which more energy is made when new bonds are made than was used to break the existing bonds are **exothermic**.

Reaction profile for an endothermic reaction

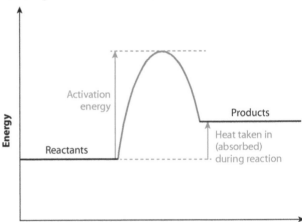

Chemical reactions in which more energy is used to break the existing bonds than is released in making the new bonds are **endothermic**.

HT The energy change of reactions

During a chemical reaction:

- bonds are broken in the reactant molecules – this is an endothermic process

- bonds are made to form the product molecules – this is an exothermic process.

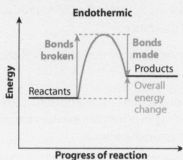

Endothermic

If more energy is used to break the bonds than is released when the bonds are made, then the reaction is endothermic.

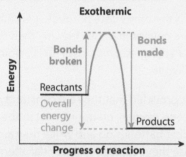

Exothermic

If more energy is released when bonds are made than is used to break the bonds, then the reaction is exothermic.

Example

Hydrogen is burned in oxygen to produce water:

hydrogen + oxygen → water

$2H_{2(g)}$ + $O_{2(g)}$ → $2H_2O_{(l)}$

2H–H + O=O → 2H–O–H

The following are **bond energies** for the **reactants** and **products**:

H–H is 436 kJ   O=O is 496 kJ   O–H is 463 kJ

Calculate the energy change.

You can calculate the energy change using this method:

1 Calculate the energy used to break bonds:

$(2 \times H–H) + O=O = (2 \times 436) + 496 = \textbf{1368 kJ}$

2 Calculate the energy released when new bonds are made:

(Water is made up of 2 × O–H bonds.)

$2 \times H–O–H = 2 \times (2 \times 463) = \textbf{1852 kJ}$

Enthalpy change (ΔH) = energy used to break bonds – energy released when new bonds are made

ΔH = 1368 – 1852

ΔH = **–484 kJ**

The reaction is **exothermic** because the energy from making the bonds is **more than** the energy needed to break the bonds.

SUMMARY

- **Exothermic reactions give out energy and the temperature of the surroundings increases; endothermic reactions take in energy and the temperature of the surroundings decreases.**

- **For a chemical reaction to occur, reacting particles must collide with sufficient energy known as the activation energy.**

QUESTIONS

QUICK TEST

1. Give one example of an exothermic reaction and one example of an endothermic reaction.

2. During an exothermic reaction, is heat given out or taken in?

QUESTIONS

EXAM PRACTICE

1. Draw and label a reaction profile for an exothermic reaction.   [3 marks]

HT 2. Calculate the energy change of the following reaction.   [3 marks]

H–H + Cl–Cl → 2H–Cl

Bond	Energy kJ/mol
H–H	436
Cl–Cl	239
H–Cl	427

Chemical cells and fuel cells

Cells and batteries

Cells contain chemicals which react to produce electricity. A typical cell consists of two different metals in contact with an electrolyte.

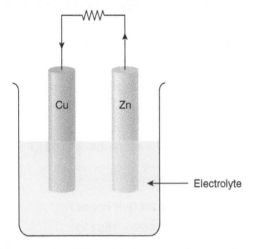

Electricity flows from the more reactive metal to the less reactive metal. In the above case, electricity would flow through the wire from the zinc to the copper.

The voltage produced by a cell depends upon many factors including…

- the type of metal used as the electrode
- the chemical used as the electrolyte.

A battery consists of two or more chemical cells connected together in series to provide a greater voltage.

In non-rechargeable cells and batteries, such as alkaline batteries, when one of the reactants has been used up the chemical reactions stop, meaning that electricity is no longer produced.

Fuel cells

Fuel cells differ from cells in that they are supplied by an external source of fuel, for example, hydrogen together with oxygen or air. The fuel is oxidised electrochemically within the cell to produce a potential difference.

The diagram at the bottom of the page shows a simple fuel cell.

The overall reaction in a hydrogen fuel cell (a fuel cell where hydrogen is the fuel) involves the oxidation of hydrogen to produce water.

$$2H_{2(g)} + O_{2(g)} \rightarrow 2H_2O_{(g)}$$

As water is the only waste product, fuel cells can be said to be more environmentally friendly than some other sources of energy (such as fossil fuels), which produce carbon dioxide.

Fuel cells, provided that they are constantly supplied with fuel and oxygen, do not run out in the same way that alkaline batteries do and while they are limited in their use they represent a potential alternative to rechargeable cells and batteries.

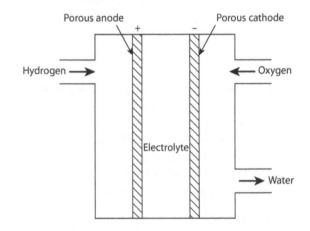

HT Half-equations for the reactions taking place in the hydrogen fuel cell

Where hydrogen is the fuel used in a fuel cell, the following reactions take place at the electrodes.

Anode: $H_{2(g)} \rightarrow 2H^+_{(aq)} + 2e^-$

Cathode: $4H^+_{(aq)} + O_{2(g)} + 4e^- \rightarrow 2H_2O_{(g)}$

Hydrogen fuel cell

Everyday chemical 'cells'

SUMMARY

● A typical cell consists of two metals in contact with an electrolyte.

● A battery consists of two or more cells connected together in series.

● Fuel cells are supplied by an external source of fuel, which oxidises electrochemically to produce a potential difference.

QUESTIONS

QUICK TEST

1. What is a fuel cell?

2. What is a battery?

HT 3. Write the half-equation for the reaction that takes place at the anode of a hydrogen fuel cell.

4. What is the waste product from a hydrogen fuel cell?

EXAM PRACTICE

1. Read the short article below.

   Cells and batteries are set to be replaced by fuel cells in the future. Fuel cells, such as hydrogen fuel cells, do not run out in the same way that cells/batteries do. Hydrogen fuel cells are also better for the environment as they do not produce any pollutant gases.

   a) Why do fuel cells not run out in the same way that cells/batteries do? **[1 mark]**

   b) Name the waste product of a hydrogen fuel cell? **[1 mark]**

   c) Suggest one challenge presented by the use of hydrogen in fuel cells. **[1 mark]**

   HT d) Write an equation for the reaction occurring at the cathode of a hydrogen fuel cell. **[1 mark]**

Rates of reaction

WS Calculating rates of reactions

The rate of a chemical reaction can be determined by measuring the quantity of a reactant used or (more commonly) the quantity of a product formed over time.

$$\text{mean rate of reaction} = \frac{\text{quantity of reactant used}}{\text{time taken}}$$

$$\text{mean rate of reaction} = \frac{\text{quantity of product formed}}{\text{time taken}}$$

For example, if 46 cm³ of gas is produced in 23 seconds then the mean rate of reaction is 2 cm³/s.

HT Rates of reaction can also be determined from graphs.

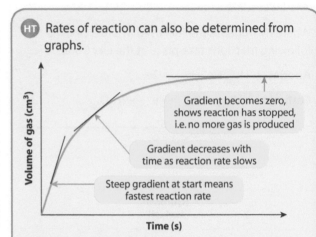

Gradient becomes zero, shows reaction has stopped, i.e. no more gas is produced

Gradient decreases with time as reaction rate slows

Steep gradient at start means fastest reaction rate

If the gradient of the tangent is calculated, this gives a numerical measure of the rate of reaction, e.g. to calculate the rate of reaction after 60 seconds draw a tangent to the curve at 60 seconds and calculate the gradient of this tangent.

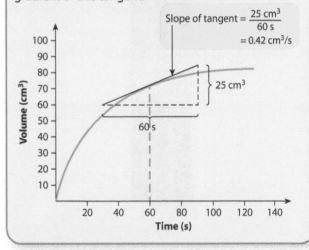

$$\text{Slope of tangent} = \frac{25\ cm^3}{60\ s}$$
$$= 0.42\ cm^3/s$$

25 cm³

60 s

Factors affecting the rates of reactions

There are five factors that affect the rate of chemical reactions:

- concentrations of the reactants in solution
- pressure of reacting gases
- surface area of any solid reactants
- temperature
- presence of a catalyst.

During experiments the rate of a chemical reaction can be found by…

- measuring the mass of the reaction mixture (e.g. if a gas is lost during a reaction)
- measuring the volume of gas produced
- observing a solution becoming opaque or changing colour.

Weighing the reaction mixture

Measuring the volume of gas produced

Observing the formation of a precipitate

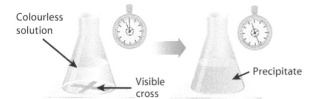

Colourless solution

Visible cross

Precipitate

QUESTIONS

QUICK TEST

1. What is the mean rate of the reaction in which 30 g of reactant is used up over 10 seconds?

2. State two factors that affect the rate of reaction.

EXAM PRACTICE

1. 0.5g of magnesium ribbon was added to a conical flask containing hydrochloric acid of concentration labelled as 100%. A diagram of the apparatus used is shown below.

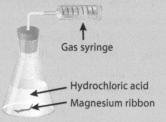

Gas syringe

Hydrochloric acid

Magnesium ribbon

The student found that it took 42 seconds to collect 90 cm^3 of hydrogen gas.

a) Calculate the mean rate of this reaction in cm^3/s.

Give your answer to 1 decimal place. **[2 marks]**

b) Suggest how the rate of reaction would change if hydrochloric acid of 50% concentration was used. **[1 mark]**

c) Other than changing concentration, suggest one other way that the rate of this reaction can be increased. **[1 mark]**

Collision theory, activation energy and catalysts

Collision theory and factors affecting rates of reaction

Collision theory explains how various factors affect rates of reaction.

It states that, for a chemical reaction to occur…

1 The reactant particles must collide with each other.

AND

2 They must collide with sufficient energy – this amount of energy is known as the activation energy.

Surface area	Temperature	Pressure	Concentration
A smaller particle size means a higher surface area to volume ratio. With smaller particles, more collisions can take place, meaning a greater rate of reaction.	Increasing the temperature increases the rate of reaction because the particles are moving more quickly and so will collide more often. Also, more particles will possess the activation energy, so a greater proportion of collisions will result in a reaction.	At a higher pressure, the gas particles are closer together, so there is a greater chance of them colliding, resulting in a higher rate of reaction.	At a higher concentration, there are more reactant particles in the same volume of solution, which increases the chance of collisions and increases the rate of reaction.
Large pieces – small surface area to volume ratio **Small pieces** – large surface area to volume ratio	**Low temperature** **High temperature**	**Low pressure** **High pressure**	**Low concentration** **High concentration**

Catalysts

Catalysts are chemicals that change the rate of chemical reactions but are not used up during the reaction. Different chemical reactions need different catalysts. In biological systems enzymes act as catalysts.

Catalysts work by providing an alternative reaction pathway of lower activation energy. This can be shown on a reaction profile.

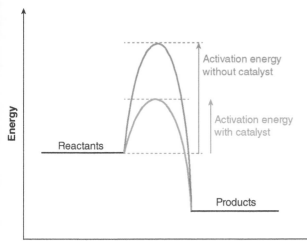

Catalysts are not reactants and so they are not included in the chemical equation.

Enzyme (biological catalyst)

Part of car catalytic converter

Zinc catalyst

SUMMARY

- Collision theory states that for a chemical reaction to occur, reactant particles must collide with each other and must collide with sufficient energy – activation energy.
- Catalysts are chemicals that change the rate of reactions but are not used up during the reaction.

QUESTIONS

QUICK TEST

1. What is meant by the term 'activation energy'?

2. Why does a higher concentration of solution increase the rate of a chemical reaction?

3. Describe collision theory.

4. How do catalysts increase the rate of chemical reactions?

EXAM PRACTICE

1. Sulfur trioxide gas can be made by heating sulfur dioxide gas with oxygen in the presence of a vanadium pentoxide, V_2O_5 catalyst.

$$2SO_{2\,(g)} + O_{2\,(g)} \rightarrow 2SO_{3\,(g)}$$

   a) Explain why heating the mixture of gases increases the rate of reaction. **[3 marks]**

   b) Explain how the vanadium pentoxide catalyst increases the rate of reaction. **[2 marks]**

   c) Increasing the pressure also increases the rate of reaction.

      Explain how. **[2 marks]**

Reversible reactions and equilibrium

Reversible reactions

In some reactions, the products of the reaction can react to produce the original reactants. These reactions are called reversible reactions.

For example:

Heating ammonium chloride:

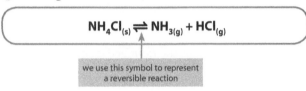

$$NH_4Cl_{(s)} \rightleftharpoons NH_{3(g)} + HCl_{(g)}$$

we use this symbol to represent a reversible reaction

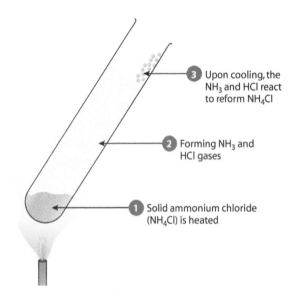

3 Upon cooling, the NH_3 and HCl react to reform NH_4Cl

2 Forming NH_3 and HCl gases

1 Solid ammonium chloride (NH_4Cl) is heated

Heating hydrated copper(II) sulfate:

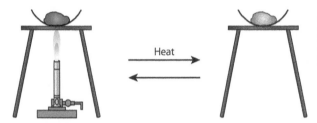

Heat
$$CuSO_4 \cdot 5H_2O_{(s)} \rightleftharpoons CuSO_{4(s)} + 5H_2O_{(l)}$$
blue   white

Heat

If the forward reaction is endothermic (absorbs heat) then the reverse reaction must be exothermic (releases heat).

Equilibrium

When a reversible reaction is carried out in a closed system (nothing enters or leaves) and the rate of the forward reaction is equal to the rate of the reverse reaction, the reaction is said to have reached equilibrium.

HT Equilibrium conditions

The relative amounts of reactants and products at equilibrium depend on the **reaction conditions**. The effect of changing conditions on reactions at equilibrium can be predicted by Le Chatelier's principle, which states that 'for a reversible reaction, if changes are made to the concentration, temperature or pressure (for gaseous reactions) then the system responds to counteract the change'.

Changing concentration

If the concentration of one of the reactants is increased, more products will be formed (to use up the extra reactant) until equilibrium is established again (i.e. the equilibrium moves to the right until a new equilibrium is established). Similarly, if the concentration of one of the products is increased, more reactants will be formed (i.e. the equilibrium moves to the left until a new equilibrium is established).

Changing temperature

Forward reaction	Effect of increasing temperature	Effect of decreasing temperature
Endothermic	Equilibrium moves to right-hand side (i.e. forward reaction)	Equilibrium moves to left-hand side (i.e. reverse reaction)
Exothermic	Equilibrium moves to left-hand side	Equilibrium moves to right-hand side

Changing pressure

In order to predict the effect of changing pressure, the number of molecules of gas on each side of the equation needs to be known:

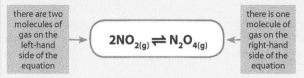

there are two molecules of gas on the left-hand side of the equation

$$2NO_{2(g)} \rightleftharpoons N_2O_{4(g)}$$

there is one molecule of gas on the right-hand side of the equation

If the pressure on a reaction at equilibrium is **increased**, the equilibrium shifts to the side of the equation with the **fewer** molecules of gas. In this case, increasing the pressure will shift the reaction to the right-hand side (i.e. producing more N_2O_4).

SUMMARY

● In reversible reactions, the products of the reaction can react to produce the original reactants.

● When the rate of the forward reaction is equal to the rate of the reverse reaction, the reaction has reached equilibrium.

QUESTIONS

QUICK TEST

1. The forward reaction in a reversible reaction is exothermic. What is the reverse reaction?

HT 2. If the forward reaction is endothermic, what is the effect on equilibrium of increasing temperature?

EXAM PRACTICE

1. Nitrogen and hydrogen gases react to form ammonia, NH_3. This is a reversible reaction.

   a) Write a balanced symbol equation for this reaction. State symbols are not required. **[2 marks]**

   b) When a mixture of these gases is placed in a closed system, eventually a state of equilibrium is reached.

   What is meant by the term equilibrium? **[2 marks]**

HT 2. Methane reacts with steam as shown by the equation below.

$$CH_{4(g)} + H_2O_{(g)} \rightleftharpoons 3H_{2(g)} + CO_{(g)} \ \Delta H = + 206 \ kJ/mol$$

   a) State and explain what will happen to the position of equilibrium if the pressure is increased. **[2 marks]**

   b) State and explain what will happen to the position of equilibrium if the temperature is increased. **[2 marks]**

Crude oil, hydrocarbons and alkanes

Crude oil

This process describes how crude oil is formed.

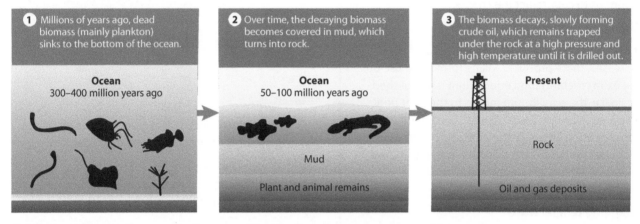

1 Millions of years ago, dead biomass (mainly plankton) sinks to the bottom of the ocean.

Ocean
300–400 million years ago

2 Over time, the decaying biomass becomes covered in mud, which turns into rock.

Ocean
50–100 million years ago

Mud

Plant and animal remains

3 The biomass decays, slowly forming crude oil, which remains trapped under the rock at a high pressure and high temperature until it is drilled out.

Present

Rock

Oil and gas deposits

As it takes so long to form crude oil, we consider it to be a finite or non-renewable resource.

Hydrocarbons and alkanes

Crude oil is a mixture of molecules called hydrocarbons.

Most hydrocarbons are members of a homologous series of molecules called alkanes.

Members of a homologous series…

- have the same general formula
- differ by CH_2 in their molecular formula from neighbouring compounds
- show a gradual trend in physical properties, e.g. boiling point
- have similar chemical properties.

Alkanes are hydrocarbons that have the general formula C_nH_{2n+2}.

Alkane	Methane, CH_4	Ethane, C_2H_6	Propane, C_3H_8	Butane, C_4H_{10}
Displayed formula	H | H − C − H | H	H H | | H − C − C − H | | H H	H H H | | | H − C − C − C − H | | | H H H	H H H H | | | | H − C − C − C − C − H | | | | H H H H

Fractional distillation

Crude oil on its own is relatively useless. It is separated into more useful components (called fractions) by fractional distillation. The larger the molecule, the stronger the intermolecular forces and so the higher the boiling point.

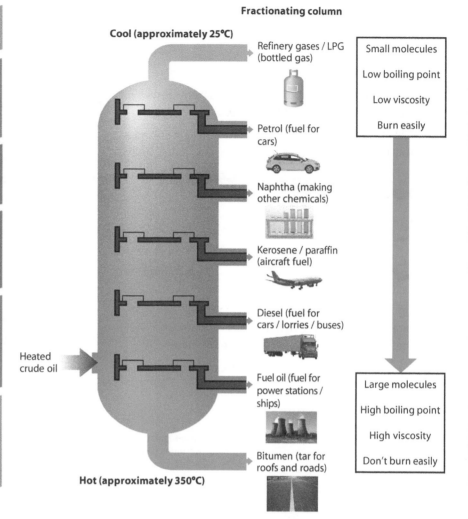

Crude oil is heated until it evaporates.

It then enters a fractionating column which is hotter at the bottom than at the top

where the molecules condense at different temperatures.

Groups of molecules with similar boiling points are collected together. They are called fractions.

The fractions are sent for processing to produce fuels and feedstock (raw materials) for the petrochemical industry

which produces many useful materials, e.g. solvents, lubricants, detergents and polymers (plastics).

Fractionating column

Cool (approximately 25°C)

Refinery gases / LPG (bottled gas)

Petrol (fuel for cars)

Naphtha (making other chemicals)

Kerosene / paraffin (aircraft fuel)

Diesel (fuel for cars / lorries / buses)

Heated crude oil

Fuel oil (fuel for power stations / ships)

Bitumen (tar for roofs and roads)

Hot (approximately 350°C)

Small molecules

Low boiling point

Low viscosity

Burn easily

Large molecules

High boiling point

High viscosity

Don't burn easily

Combustion of hydrocarbons

Some of the fractions of crude oil (e.g. petrol and kerosene) are used as fuels. Burning these fuels releases energy. Combustion (burning) reactions are oxidation reactions. When the fuel is fully combusted, the carbon in the hydrocarbons is oxidised to carbon dioxide and the hydrogen is oxidised to water. For example, the combustion of methane:

$$CH_{4(g)} + 2O_{2(g)} \rightarrow CO_{2(g)} + 2H_2O_{(l)}$$

SUMMARY

- **Crude oil is a mixture of hydrocarbons.**
- **Complete combustion of hydrocarbon forms carbon dioxide and water.**
- **Crude oil is separated into useful fractions by fractional distillation.**

QUESTIONS

QUICK TEST

1. What is crude oil formed from?

2. What is the general formula of alkanes?

3. What is the molecular formula of ethane?

EXAM PRACTICE

1. After being extracted from the Earth, crude oil first undergoes fractional distillation.

   a) Explain why crude oil is fractionally distilled. **[1 mark]**

   b) Explain how fractional distillation separates crude oil into different fractions. **[3 marks]**

Cracking and alkenes

Cracking and alkenes

Many of the long-chain hydrocarbons found in crude oil are not very useful. Cracking is the process of turning a long-chain hydrocarbon into shorter, more useful ones.

Long-chain hydrocarbon

Short-chain hydrocarbons

Cracking is done by passing hydrocarbon vapour over a hot catalyst or mixing the hydrocarbon vapour with steam before heating it to a very high temperature.

The diagram shows how cracking can be carried out in the laboratory.

Aluminium oxide or broken pot (catalyst)

Gaseous short-chain hydrocarbon molecules (alkene)

Heat

Long-chain hydrocarbon – liquid paraffin (alkane) soaked in mineral wool

Liquid short-chain hydrocarbon molecules (alkane)

Cold water

Cracking produces alkanes and alkenes. The small-molecule alkanes that are formed during cracking are in high demand as fuels. The alkenes are mostly used to make plastics by the process of polymerisation (see page 58).

There are many different equations that can represent cracking. This is because the long hydrocarbon can break in many different places. A typical equation for the cracking of the hydrocarbon decane ($C_{10}H_{22}$) is:

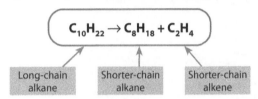

$$C_{10}H_{22} \rightarrow C_8H_{18} + C_2H_4$$

| Long-chain alkane | Shorter-chain alkane | Shorter-chain alkene |

The presence of alkenes can be detected using bromine water. Alkenes decolourise bromine water but when it is mixed with alkanes, the bromine water stays orange.

Unsaturated alkene (C=C) + bromine water

Saturated alkane (C–C) + bromine water

Alkenes

Alkenes are hydrocarbons with a carbon–carbon double bond. Molecules with carbon–carbon double bonds are described as unsaturated molecules because there are two fewer hydrogen atoms than an alkane with the same number of carbon atoms. The general formula for the alkene homologous series is C_nH_{2n}.

Alkene	Ethene, C_2H_4	Propene, C_3H_6	Butene (but-1-ene), C_4H_8	But-2-ene, C_4H_8	Pent-1-ene, C_5H_{10}
Displayed formula	$\begin{array}{c} H \quad\quad H \\ \backslash\quad/ \\ C=C \\ /\quad\backslash \\ H \quad\quad H \end{array}$	$\begin{array}{c} H \quad\quad H \\ \backslash\quad/\;\;| \\ C=C-C-H \\ /\quad\;\;|\;\;| \\ H \quad H\;H \end{array}$	$\begin{array}{c} H \quad\quad H\;H \\ \backslash\quad/\;\;|\;\;| \\ C=C-C-C-H \\ /\quad\;\;|\;\;|\;\;| \\ H \quad H\;H\;H \end{array}$	$\begin{array}{c} H\;H\;H\;H \\ |\;\;|\;\;|\;\;| \\ H-C-C=C-C-H \\ |\quad\quad\quad| \\ H \quad\quad\quad H \end{array}$	$\begin{array}{c} H \quad\quad H\;H\;H\;H \\ \backslash\quad/\;\;|\;\;|\;\;| \\ C=C-C-C-C-H \\ /\quad\;\;|\;\;|\;\;| \\ H \quad H\;H\;H \end{array}$

Alkenes usually react by atoms adding across the carbon–carbon double bond.

Reactant	Notes	Equation
Oxygen	Alkenes react with oxygen in combustion reactions. Incomplete combustion occurs when alkenes do not completely oxidise. When alkenes burn with a sooty flame incomplete combustion is taking place.	e.g. the complete combustion of ethene: $C_2H_{4(g)} + 3O_2 \rightarrow 2CO_{2(g)} + 3H_2O_{(l)}$
Hydrogen	Hydrogen atoms add across the carbon–carbon double bond so that a carbon–carbon single bond remains.	A nickel catalyst is required, e.g. when ethene reacts with hydrogen, ethane is formed: 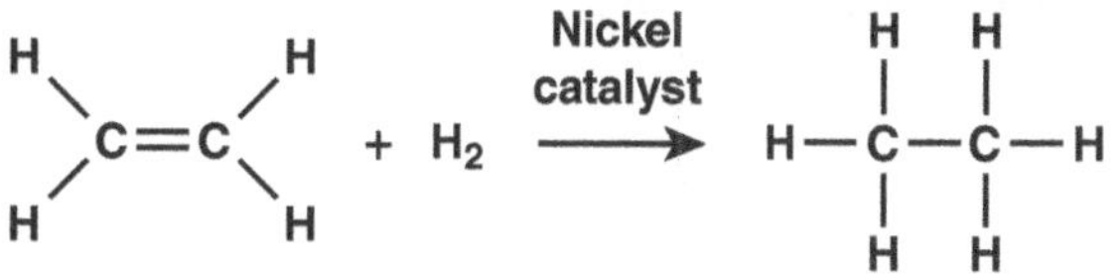
Water	One of the hydrogen atoms in water adds to one side of the carbon–carbon double bond. The remaining OH atoms add to the other carbon atom.	This reaction takes place at a high temperature so that the water is present as steam. An acid catalyst (e.g. concentrated sulfuric acid) is required: e.g. ethene reacts with steam to form the alcohol ethanol. 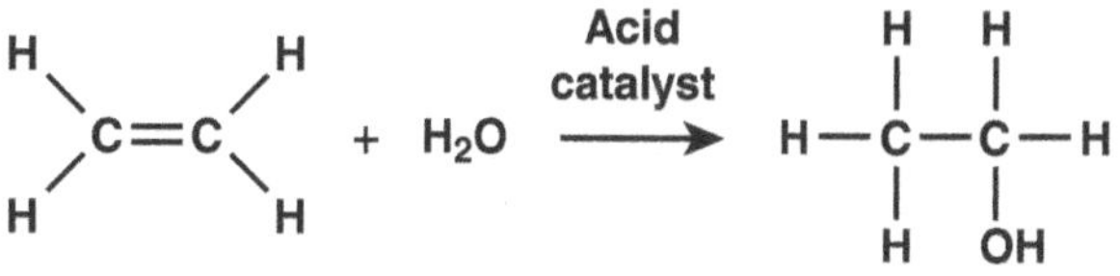
Halogens (e.g. Cl_2 / Br_2)	Halogen atoms add across the carbon–carbon double bond so that a carbon–carbon single bond remains.	Ethene reacts with bromine to form dibromoethane: 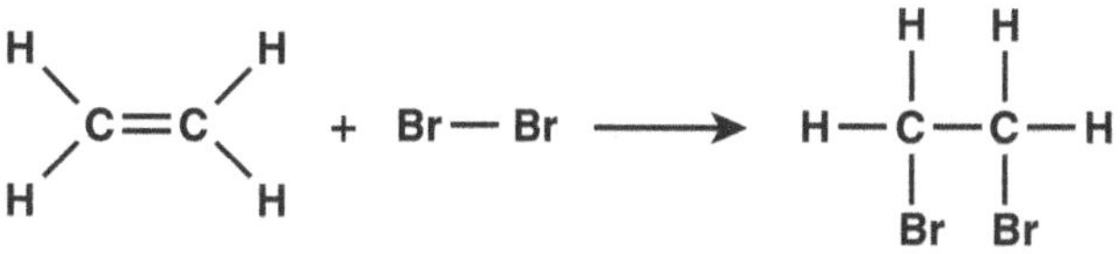

Other alkenes react in the same way as ethene. This is because all alkenes contain the same functional group, for example, the reaction of propene with bromine:

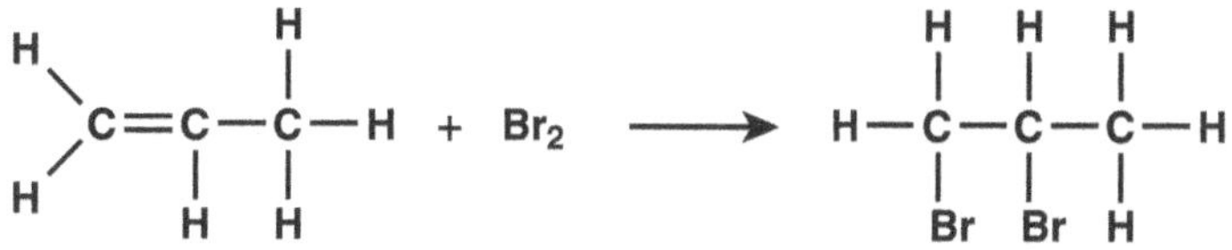

QUESTIONS

QUICK TEST

1. Draw the displayed formula of ethene.
2. What is cracking?

QUESTIONS

EXAM PRACTICE

1. When octane, C_8H_{18}, is cracked it can form a molecule of butane and two molecules of an alkene.

   a) Write a balanced symbol equation for this reaction. **[2 marks]**

   b) Describe how the presence of an alkene can be confirmed. **[2 marks]**

DAY 5 ⏱ 30 Minutes

Alcohols and carboxylic acids

Alcohols

Alcohols contain the functional group –OH. The first four members of the homologous series of alcohols are shown in the table.

Alcohol (all names end in 'ol')	Displayed formula	Structural formula (always ends in –OH)
Methanol	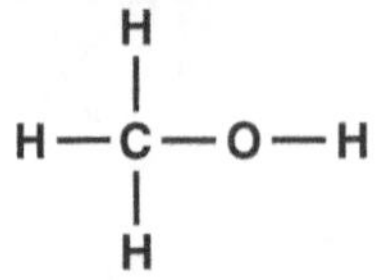 H—C—O—H (with H above and below C)	CH_3OH
Ethanol	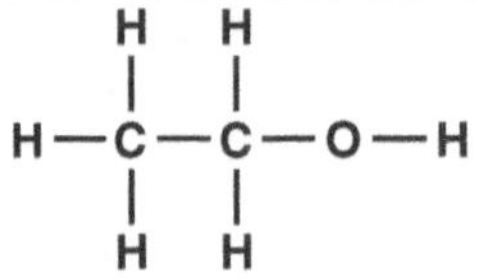 H—C—C—O—H (with H above and below each C)	CH_3CH_2OH
Propanol	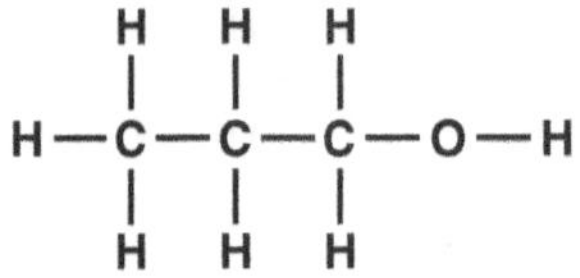 H—C—C—C—O—H (with H above and below each C)	$CH_3CH_2CH_2OH$
Butanol	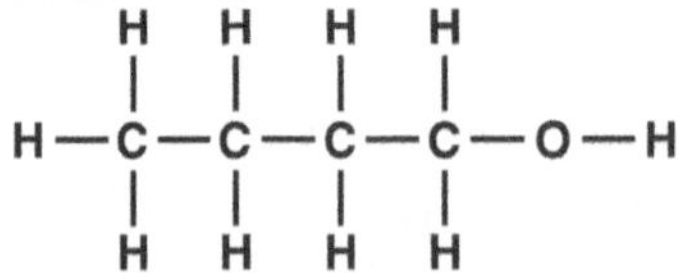 H—C—C—C—C—O—H (with H above and below each C)	$CH_3CH_2CH_2CH_2OH$

Alcohols...

- dissolve in water to form neutral solutions
- react with sodium to produce hydrogen
- burn in air to produce carbon dioxide and water

    For example: $C_2H_5OH_{(l)} + 3O_{2(g)} \rightarrow 2CO_{2(g)} + 3H_2O_{(l)}$

- can be oxidised by acidified potassium manganate(VII) to form carboxylic acids.

Aqueous solutions of ethanol are produced by the anaerobic fermentation of sugar solution by yeast at 30°C. The product can be distilled to produce a more concentrated solution of ethanol. Ethanol is the alcohol in alcoholic drinks.

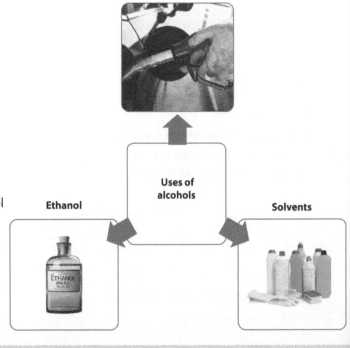

Fuels

Uses of alcohols

Ethanol

Solvents

Carboxylic acids

Carboxylic acids contain the functional group –COOH. The first four members of the homologous series of carboxylic acids are shown in the table.

Carboxylic acid (all names end in '... oic acid')	Displayed formula	Structural formula (always ends in COOH)
Methanoic acid	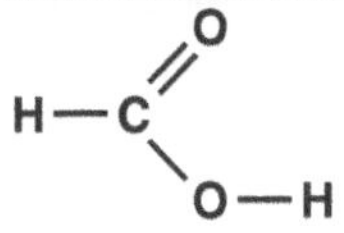 H—C with =O and O—H	HCOOH
Ethanoic acid	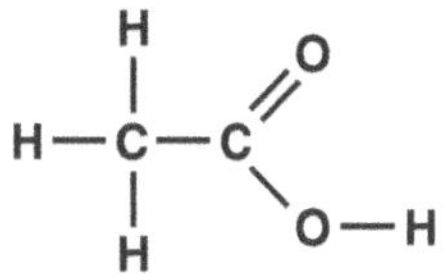 H—C(H)(H)—C with =O and O—H	CH_3COOH
Propanoic acid	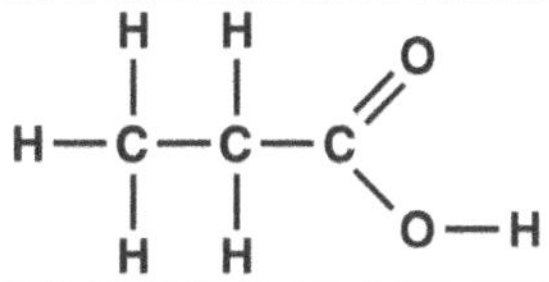 H—C(H)(H)—C(H)(H)—C with =O and O—H	CH_3CH_2COOH
Butanoic acid	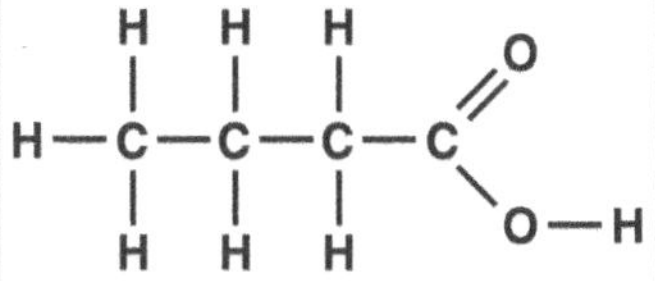 H—C(H)(H)—C(H)(H)—C(H)(H)—C with =O and O—H	$CH_3CH_2CH_2COOH$

Carboxylic acids...

● Dissolve in water to form acidic solutions

● React with carbonates to form carbon dioxide

> HT ● Are weak acids, i.e. they do not completely ionise when dissolved in water

● React with alcohols in the presence of an acid catalyst to produce esters.

For example: ethanoic acid reacts with ethanol to produce water and the ester ethyl ethanoate.

Ethyl ethanoate

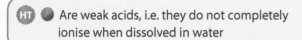

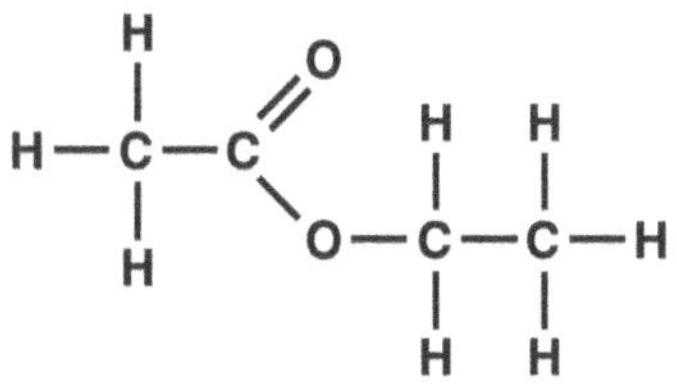

QUESTIONS

QUICK TEST

1. What is the structural formula of ethanol?

2. Which alcohol is present in alcoholic drinks?

3. Draw the displayed formula of methanoic acid.

4. When ethanoic acid reacts with ethanol, what is formed?

EXAM PRACTICE

1. a) Name the alcohol shown below. [1 mark]

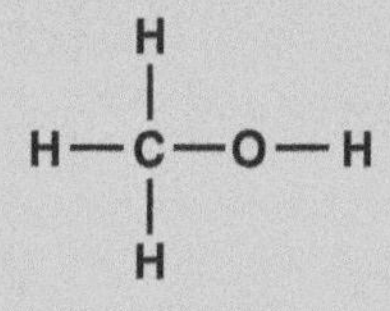

   b) Name the gas produced when sodium is added to the above alcohol. [1 mark]

   c) What is produced when acidified potassium manganate(VII) is added to the above alcohol? [1 mark]

2. a) What type of substance needs to be added to propanoic acid in order to make an ester? [1 mark]

   HT b) Propanoic acid is a weak acid.

   Explain the meaning of the term 'weak' when referring to acids. [2 marks]

Polymerisation & natural polymers

Addition polymerisation

Addition polymerisation is the process of joining lots of small molecules (monomers) together to form very large molecules (polymers). The monomers must be alkenes because a carbon–carbon double bond is needed for addition polymerisation to occur.

Ethene monomers (unsaturated)

...and thousands more...

Poly(ethene) polymers (saturated)

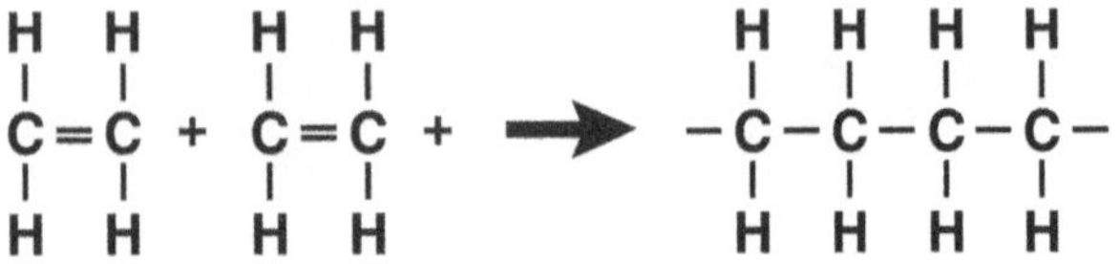

...and on and on…

The general equation below can be used to represent the formation of any addition polymer.

Repeat unit

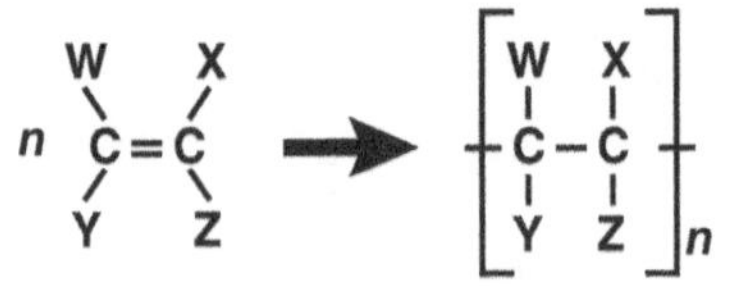

Monomer – any alkene   Polymer

The repeat unit in an addition polymer has the same atoms as the monomer because there are no other molecules formed in the reaction.

		Polymer		
Monomer	Repeat unit	Name	Uses	Properties
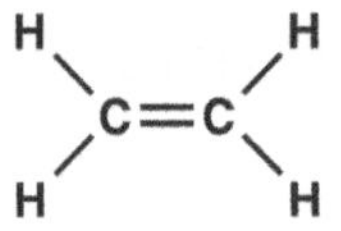	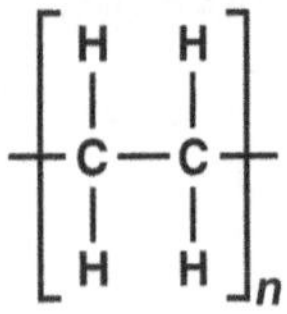	Poly(ethene)	Plastic shopping bags Water bottles	Strong and light
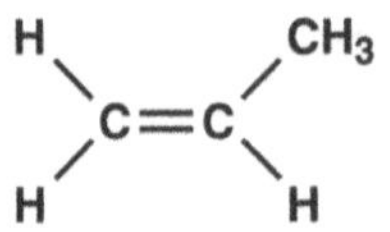	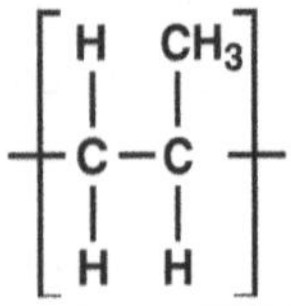	Poly(propene)	Crates and ropes	Strong and rigid
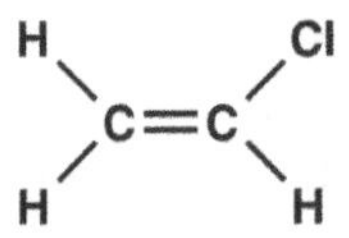	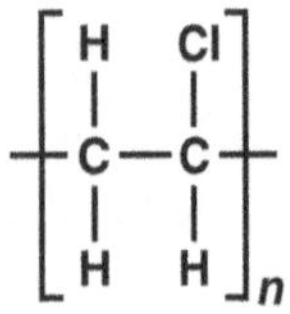	Poly(chloroethene) (PVC)	Water pipes and electrical cable insulation	Waterproof Electrical insulator Flexible
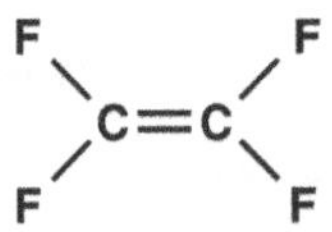	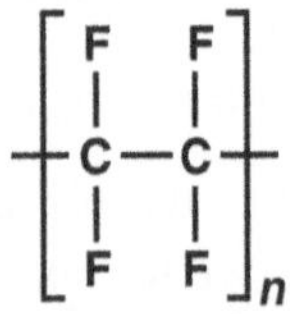	Poly(tetrafluoroethene) (PTFE)	Lining for non-stick saucepans	Slippery Chemically inert

🄗 Condensation polymerisation

Condensation polymerisation takes place when the monomers have two functional groups in the molecule, such as ethane diol, which contains an alcohol (–OH) functional group at each end of the molecule.

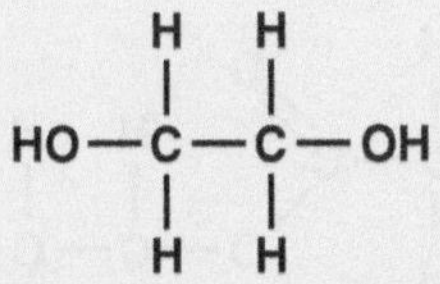

HT Hexanedioic acid is a molecule that contains a carboxylic acid group (–COOH) at each end of the molecule.

$$HO-\overset{\overset{\displaystyle O}{\|}}{C}-(CH_2)_4-\overset{\overset{\displaystyle O}{\|}}{C}-OH$$

Alcohols react with carboxylic acids to form esters (see page 57). These two molecules polymerise to form a polyester. In condensation polymerisation, a small molecule (usually water) is also produced.

$$n\ HO-\square-OH\ +\ n\ \overset{\overset{\displaystyle O}{\|}}{C}-\square-\overset{\overset{\displaystyle O}{\|}}{C}_{OH}$$

Ethane diol   Hexanedioic acid

$$\Bigg[O-\square-O-\overset{\overset{\displaystyle O}{\|}}{C}-\square-\overset{\overset{\displaystyle O}{\|}}{C}\Bigg]_n\ +\ (2n-1)\ H_2O$$

In this diagram, blocks □ are used to represent the carbon chains in each monomer molecule (i.e. CH_2CH_2 and $CH_2CH_2CH_2CH_2$), as these are not involved in the polymerisation reaction.

Amino acids

Amino acids have two different functional groups in a molecule. Amino acid molecules react with each other to form polypeptides. Different amino acids can be combined in the same chain to form proteins.

Glycine is an amino acid with the formula H_2NCH_2COOH.

$$\underset{H}{\overset{H}{N}}-\underset{H}{\overset{H}{C}}-\overset{\overset{\displaystyle O}{\|}}{C}-O-H$$

Glycine undergoes condensation polymerisation (the NH_2 functional group in one molecule reacts with the –COOH functional group in another) to form the polypeptide $(-HNCH_2COO-)_n$

$$n\ H_2NCH_2COOH \rightarrow (-HNCH_2COO-)_n + nH_2O$$

DNA

DNA (deoxyribonucleic acid) is a large molecule that is essential for life. DNA encodes genetic instructions for the development and functioning of living organisms and viruses.

Most DNA molecules are two polymer chains, made from four different monomers called nucleotides. These are arranged in the form of a double helix.

DNA double helix

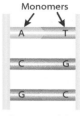

A section of the double helix

Other naturally occurring polymers

There are many polymers found in nature that are important for life. For example:

● proteins
● starch and cellulose

Starch and cellulose are polymers of sugars. Sugars, starch and cellulose are carbohydrates.

SUMMARY

● Addition polymerisation is the process of joining monomers to form polymers; the monomers must be alkenes.
● **HT** Condensation polymerisation occurs when the monomer has two functional groups in the molecule.
● Most DNA molecules are two polymer chains, made from four monomers called nucleotides.

QUESTIONS

QUICK TEST

1. Suggest a use for poly(ethene).
2. Name two naturally occurring polymers.

EXAM PRACTICE

1. Polymerisation is the process of joining together lots of small molecules to make long chain molecules. Propene undergoes addition polymerisation to form the polymer poly(propene).

   a) What feature of propene means that it will undergo addition polymerisation? **[1 mark]**

   b) Name the other type of polymerisation. **[1 mark]**

Purity, formulations and chromatography

Purity

In everyday language, a pure substance can mean a substance that has had nothing added to it (i.e. in its natural state), such as milk.

In chemistry, a pure substance is a single element or compound (i.e. not mixed with any other substance).

Pure elements and compounds melt and boil at specific temperatures. For example, pure water freezes at 0°C and boils at 100°C. However, if something is added to water (e.g. salt) then the freezing point decreases (i.e. goes below 0°C) and the boiling point rises above 100°C.

Formulations

A formulation is a mixture that has been designed as a useful product. Many formulations are complex mixtures in which each ingredient has a specific purpose.

Formulations are made by mixing the individual components in carefully measured quantities to ensure that the product has the correct properties.

Chromatography

Chromatography is used to separate mixtures of dyes. It is used to help identify substances.

In paper chromatography, a solvent (the mobile phase) moves up the paper (the stationary phase) carrying different components of the mixture different distances, depending on their attraction for the paper and the solvent. In thin layer chromatography (TLC), the stationary phase is a thin layer of an inert substance (e.g. silica) supported on a flat, unreactive surface (e.g. a glass plate).

In the chromatogram below, substance X is being analysed and compared with samples A, B, C, D and E.

It can be seen from the chromatogram that substance X has the same pattern of spots as sample D. This means that sample X and sample D are the same substance. Pure compounds (e.g. compound A) will only produce one spot on a chromatogram.

Examples of formulations

Fuels

Cleaning materials

Paints

Medicines

Foods

Fertilisers

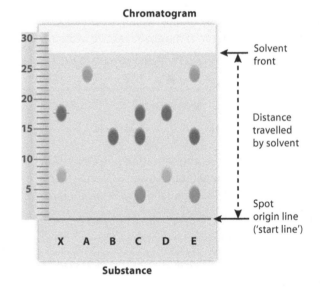

Chromatogram

Solvent front

Distance travelled by solvent

Spot origin line ('start line')

X A B C D E

Substance

The ratio of the distance moved by the compound to the distance moved by the solvent is known as its R_f value.

$$R_f = \frac{\text{distance moved by substance}}{\text{distance moved by solvent}}$$

Different compounds have different R_f values in different solvents. This can be used to help identify unknown compounds by comparing R_f values with known substances.

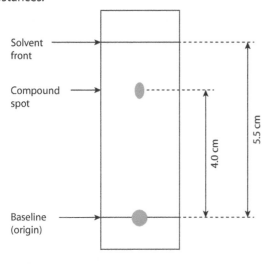

Solvent front

Compound spot

5.5 cm

4.0 cm

Baseline (origin)

In this case, the R_f value is 0.73 $\left(\frac{4.0}{5.5}\right)$

In **gas chromatography** (GC), the mobile phase is an inert gas (e.g. helium). The stationary phase is a very thin layer of an inert liquid on a solid support, such as beads of silica packed into a long thin tube. GC is a more sensitive method than TLC for separating mixtures, and it allows you to determine the amount of each chemical in the mixture.

SUMMARY

● A pure substance is a single element or compound (i.e. not mixed with any other substance).

● A formulation is a mixture of substances, made to be a useful product.

● Chromatography is a process used to separate mixtures of dyes and help identify substances.

QUESTIONS

QUICK TEST

1. What is chromatography used for?

2. What is a formulation?

3. Give two examples of substances that are formulations.

4. In a chromatogram, the solvent travelled 4 cm and a dye travelled 2.8 cm. What is the R_f value of the dye?

EXAM PRACTICE

1. Paper chromatography was carried out to investigate the dyes present in four different coloured inks. The chromatogram is shown below.

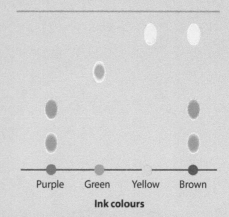

Purple   Green   Yellow   Brown

Ink colours

a) Which two inks only contain one dye?

[1 mark]

b) How can the brown ink be made from the other inks? [2 marks]

Identification of gases

Testing for hydrogen

Hydrogen burns with a squeaky pop when tested with a lighted splint.

Pop!

Test tube of hydrogen

Lighted splint

Testing for oxygen

Oxygen relights a glowing splint.

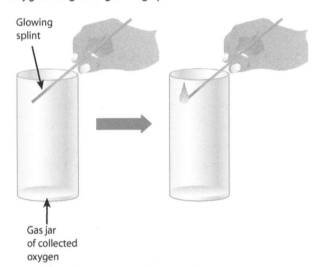

Glowing splint

Gas jar of collected oxygen

O_2

Oxygen

H_2

Hydrogen

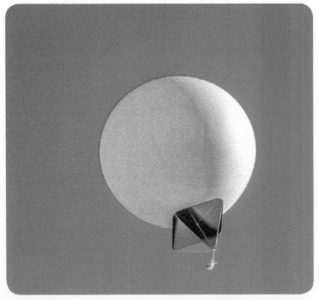

A hydrogen-filled weather balloon

Oxygen is used in breathing masks

Testing for carbon dioxide

When carbon dioxide is mixed with or bubbled through limewater (calcium hydroxide solution) the limewater turns milky (cloudy).

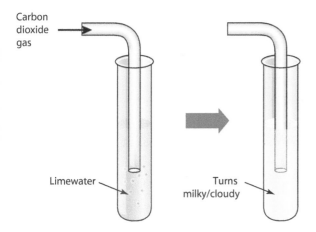

Carbon dioxide gas

Limewater

Turns milky/cloudy

Testing for chlorine

Chlorine turns moist blue litmus paper red before bleaching it and turning it white.

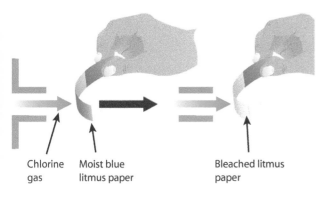

Chlorine gas

Moist blue litmus paper

Bleached litmus paper

Chlorine gas is very toxic

Carbon dioxide is released when fossil fuels are burnt

SUMMARY

- A splint can be used to test for hydrogen and oxygen; test for hydrogen using a lighted splint and test for oxygen using a glowing splint.
- Limewater is used to test for carbon dioxide.
- Litmus paper is used to test for chlorine.

QUESTIONS

QUICK TEST

1. Which gas relights a glowing splint?
2. What is the test for hydrogen gas?
3. What is calcium hydroxide solution also known as?
4. What happens to limewater when it is mixed with carbon dioxide gas?

EXAM PRACTICE

1. When potassium manganate is added to concentrated hydrochloric acid, a gas that is suspected to be chlorine is produced.

   Describe how to confirm that the gas is chlorine. **[3 marks]**

Identification of cations

Flame tests

Flame tests can be used to identify metal ions.

Lithium, sodium, potassium, calcium and copper compounds can be recognised by the distinctive colours they produce in a flame test.

To do a flame test, follow this method:

1 Heat and then dip a piece of nichrome (a nickel–chromium alloy) wire in concentrated hydrochloric acid to clean it.

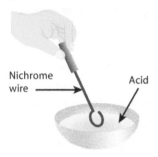

Nichrome wire → ← Acid

2 Dip the wire in the compound.

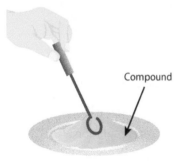

Compound →

3 Put it into a Bunsen flame. Different colours will indicate the presence of certain ions.

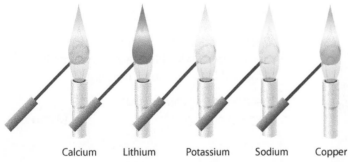

Calcium   Lithium   Potassium   Sodium   Copper

The distinctive colours are:

● orange-red / brick red for **calcium** (Ca^{2+})

● crimson red for **lithium** (Li^+)

● lilac for **potassium** (K^+)

● yellow for **sodium** (Na^+)

● blue / green for **copper** (Cu^{2+}).

If a sample containing a mixture of ions is used then some flame colours can be masked.

Using sodium hydroxide

Metal compounds in solution contain **metal ions**. Some of these form precipitates – **insoluble** solids that come out of solution when sodium hydroxide solution is added to them.

For example, when sodium hydroxide solution is added to iron(III) chloride solution, a white precipitate of iron(III) hydroxide is formed (as well as sodium chloride solution). You can see how this precipitate is formed by considering the ions involved.

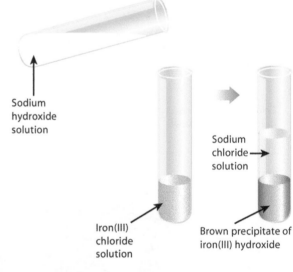

Sodium hydroxide solution

Sodium chloride solution

Iron(III) chloride solution

Brown precipitate of iron(III) hydroxide

The precipitates formed when metal ions are mixed with sodium hydroxide solution			
Metal ion	Precipitate formed		
	Precipitate	Precipitate colour	Equation for formation of precipitate
Aluminium $Al^{3+}_{(aq)}$	Aluminium hydroxide	White (dissolves with excess sodium hydroxide)	$Al^{3+}_{(aq)} + 3OH^-_{(aq)} \rightarrow Al(OH)_{3(s)}$
Calcium $Ca^{2+}_{(aq)}$	Calcium hydroxide	White	$Ca^{2+}_{(aq)} + 2OH^-_{(aq)} \rightarrow Ca(OH)_{2(s)}$
Magnesium $Mg^{2+}_{(aq)}$	Magnesium hydroxide	White	$Mg^{2+}_{(aq)} + 2OH^-_{(aq)} \rightarrow Mg(OH)_{2(s)}$
Copper(II) $Cu^{2+}_{(aq)}$	Copper(II) hydroxide	Blue	$Cu^{2+}_{(aq)} + 2OH^-_{(aq)} \rightarrow Cu(OH)_{2(s)}$
Iron(II) $Fe^{2+}_{(aq)}$	Iron(II) hydroxide	Green	$Fe^{2+}_{(aq)} + 2OH^-_{(aq)} \rightarrow Fe(OH)_{2(s)}$
Iron(III) $Fe^{3+}_{(aq)}$	Iron(III) hydroxide	Brown	$Fe^{3+}_{(aq)} + 3OH^-_{(aq)} \rightarrow Fe(OH)_{3(s)}$

SUMMARY

● Flame tests can be used to identify metal ions. When put in a Bunsen flame, different colours will indicate the presence of certain ions.

● Sodium hydroxide solution can be added to solutions to indicate whether metal ions are present. This is confirmed by observing whether any precipitates are formed.

QUESTIONS

QUICK TEST

1. What colour flame is seen when a flame test is done on a compound containing potassium?

2. Which metal gives a yellow flame in a flame test?

3. What is the name of the precipitate formed when a solution containing aluminium ions reacts with sodium hydroxide solution?

4. Write an equation for the formation of the precipitate formed when a solution containing Fe^{2+} ions reacts with sodium hydroxide.

QUESTIONS

EXAM PRACTICE

1. Table salt is sodium chloride. An alternative to table salt, 'Low Sodium Salt', contains potassium chloride mixed with sodium chloride.

   a) Describe how a flame test can confirm the presence of sodium in table salt. **[1 mark]**

   b) Explain why a flame test carried out on 'Low Sodium Salt' would not be able to confirm the presence of potassium. **[2 marks]**

2. Aluminium chloride and magnesium chloride are both white soluble solids.

   Explain how a sample of aluminium chloride can be distinguished from a sample of magnesium chloride. **[4 marks]**

Identification of anions

Compounds containing the carbonate (CO_3^{2-}) ion

An unknown solid can be tested with dilute acid to see if it contains carbonate ions, CO_3^{2-}. If the solid is a carbonate, it will react with the acid to form a salt, water and carbon dioxide gas, which fizzes. For example:

calcium carbonate + hydrochloric acid → calcium chloride + carbon dioxide + water
$CaCO_{3(s)}$ + $2HCl_{(aq)}$ → $CaCl_{2(aq)}$ + $CO_{2(g)}$ + $H_2O_{(l)}$

Limewater can be used to test for carbon dioxide. If the gas is present, the limewater will turn milky / cloudy.

Most carbonate compounds are insoluble but some (e.g. sodium carbonate and potassium carbonate) are soluble in water and produce solutions containing carbonate ions.

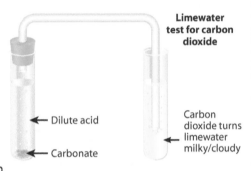

Limewater test for carbon dioxide

← Dilute acid

← Carbonate

Carbon dioxide turns limewater milky/cloudy

Identifying dissolved ions

The **dissolved ions** of some salts are easy to identify as they will undergo **precipitation** reactions.

Sulfates (SO_4^{2-}) can be detected using dilute hydrochloric acid and barium chloride solution.

A white precipitate of barium sulfate forms, as in the following example.

sodium sulfate + barium chloride → barium sulfate + sodium chloride
$Na_2SO_{4(aq)}$ + $BaCl_{2(aq)}$ → $BaSO_{4(s)}$ + $2NaCl_{(aq)}$
(white)

Silver nitrate solution, in the presence of dilute nitric acid, is used to detect halide ions.

Halides are the ions made by the halogens (group 7).

With silver nitrate:

- chlorides (Cl^-) form a white precipitate
- bromides (Br^-) form a cream precipitate
- iodides (I^-) form a pale yellow precipitate.

sodium chloride + silver nitrate → silver chloride + sodium nitrate
$NaCl_{(aq)}$ + $AgNO_{3(aq)}$ → $AgCl_{(s)}$ + $NaNO_{3(aq)}$
(white)

sodium bromide + silver nitrate → silver bromide + sodium nitrate
$NaBr_{(aq)}$ + $AgNO_{3(aq)}$ → $AgBr_{(s)}$ + $NaNO_{3(aq)}$
(cream)

sodium bromide + silver nitrate → silver iodide + sodium nitrate
$NaI_{(aq)}$ + $AgNO_{3(aq)}$ → $AgI_{(s)}$ + $NaNO_{3(aq)}$
(pale yellow)

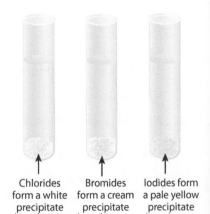

Chlorides form a white precipitate

Bromides form a cream precipitate

Iodides form a pale yellow precipitate

Instrumental methods of analysis

As well as chemical tests, elements and compounds can also be analysed and identified by instrumental methods.

Advantages of instrumental methods compared with chemical tests		
● More accurate	● More sensitive	● Can be used with very small samples

Flame emission spectroscopy

Flame emission spectroscopy can be used to identify metal ions present in solutions.

1. The sample is put into a flame.
2. The light given out is passed through a spectroscope/flame photometer.
3. A line spectrum is produced.
4. This is compared with reference spectra.
5. Metal ions present are identified.
6. The concentration of ions in solution can be determined by reference to a calibration curve.

The line spectra of some metals

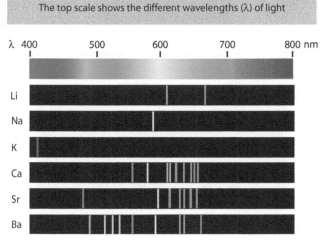

The top scale shows the different wavelengths (λ) of light

λ 400   500   600   700   800 nm

Li

Na

K

Ca

Sr

Ba

SUMMARY

● Dilute acid can be used to test an unknown solid for carbonate ions.

● The dissolved ions of some salts can be easily identified as they undergo precipitation reactions.

● Compared with chemical tests, instrumental tests are more rapid, more sensitive and can be used with very small samples.

● Metal ions in solutions can be identified using flame emission spectroscopy.

QUESTIONS

QUICK TEST

1. Which ion reacts with acid to form CO_2 gas?
2. What colour precipitate is formed when iodide ions react with silver nitrate solution?

EXAM PRACTICE

1. Describe how the presence of bromide ions in a sample of solid potassium bromide can be confirmed. **[3 marks]**

2. Flame emission spectroscopy can be used to determine the presence of metal ions in a compound.

   Give two advantages of an instrumental method of analysis, such as flame emission spectroscopy, over flame tests carried out in a laboratory. **[2 marks]**

Evolution of the atmosphere

The Earth's early atmosphere

Theories about the composition of Earth's early atmosphere and how the atmosphere was formed have changed over time. Evidence for the early atmosphere is limited because the Earth is approximately 4.6 billion years old and humans were not around to record data.

The table below gives one theory to explain the evolution of the atmosphere.

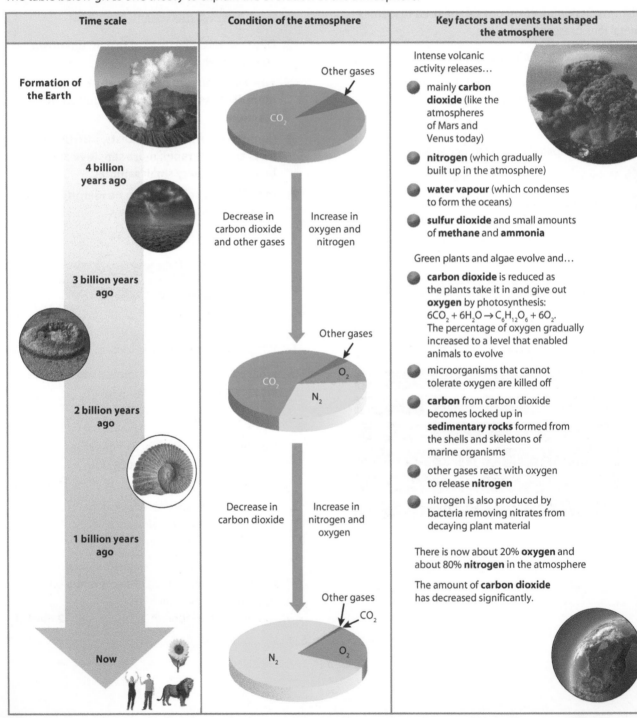

Time scale	Condition of the atmosphere	Key factors and events that shaped the atmosphere
Formation of the Earth	Other gases / CO_2	Intense volcanic activity releases… • mainly **carbon dioxide** (like the atmospheres of Mars and Venus today) • **nitrogen** (which gradually built up in the atmosphere) • **water vapour** (which condenses to form the oceans) • **sulfur dioxide** and small amounts of **methane** and **ammonia**
4 billion years ago	Decrease in carbon dioxide and other gases / Increase in oxygen and nitrogen	Green plants and algae evolve and… • **carbon dioxide** is reduced as the plants take it in and give out **oxygen** by photosynthesis: $6CO_2 + 6H_2O \rightarrow C_6H_{12}O_6 + 6O_2$. The percentage of oxygen gradually increased to a level that enabled animals to evolve
3 billion years ago	Other gases / O_2 / CO_2 / N_2	• microorganisms that cannot tolerate oxygen are killed off • **carbon** from carbon dioxide becomes locked up in **sedimentary rocks** formed from the shells and skeletons of marine organisms
2 billion years ago	Decrease in carbon dioxide / Increase in nitrogen and oxygen	• other gases react with oxygen to release **nitrogen** • nitrogen is also produced by bacteria removing nitrates from decaying plant material
1 billion years ago	Other gases / CO_2 / O_2 / N_2	There is now about 20% **oxygen** and about 80% **nitrogen** in the atmosphere The amount of **carbon dioxide** has decreased significantly.
Now		

Composition of the atmosphere today

The proportions of gases in the atmosphere have been more or less the same for about 200 million years. **Water vapour** may also be present in varying quantities (0–3%).

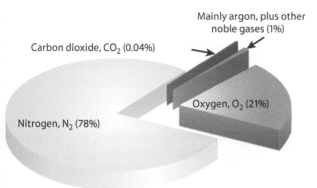

Carbon dioxide, CO_2 (0.04%)

Mainly argon, plus other noble gases (1%)

Oxygen, O_2 (21%)

Nitrogen, N_2 (78%)

How carbon dioxide decreased

The amount of carbon dioxide in the atmosphere today is much less than it was when the atmosphere first formed. This is because…

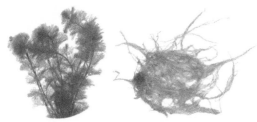

● green plants and algae use carbon dioxide for photosynthesis

● carbon dioxide is used to form sedimentary rocks, e.g. limestone

● fossil fuels such as oil (see page 52) and coal (a sedimentary rock made from thick plant deposits that were buried and compressed at high temperatures over millions of years) have captured CO_2.

SUMMARY

● Although theories differ, it is thought the Earth was formed 4.6 billion years ago and the atmosphere was mostly carbon dioxide.

● The composition of the atmosphere has changed over time and is now mostly nitrogen.

● Carbon dioxide decreased due to it being used in photosynthesis, forming sedimentary rocks and CO_2 being captured as fossil fuels.

QUESTIONS

QUICK TEST

1. Explain how oxygen became present in the atmosphere.

2. Name two gases that volcanoes released into the early atmosphere.

3. How much of the atmosphere today is water vapour?

4. What gas do green plants and algae use for photosynthesis?

EXAM PRACTICE

1. Describe and explain the changes that green plants and algae have had on the composition of the atmosphere since their evolution. **[3 marks]**

2. A sample of air was found to have the following composition.

   Total volume = 820 cm³

   Volume of oxygen = 164 cm³
   Volume of nitrogen = 640 cm³
   Volume of other gases =

   a) Calculate the volume of other gases in the sample of air. **[1 mark]**

   b) Name two gases that are present in the other gases. **[2 marks]**

   c) Calculate the percentage of oxygen in the sample of air. **[1 mark]**

Climate change

WS Human activity and global warming

Some human activities increase the amounts of greenhouse gases in the atmosphere including…

Combustion of fossil fuels releases carbon dioxide into the atmosphere

Increased animal farming releases more methane into the atmosphere, e.g. as a by-product of digestion and decomposition of waste

Deforestation reduces the amount of carbon dioxide removed from the atmosphere by photosynthesis

Decomposition of rubbish in landfill sites also releases methane into the atmosphere

Greenhouse gases

The temperature on Earth is maintained at a level to support life by the greenhouse gases in the atmosphere. These gases allow short wavelength radiation from the Sun to pass through but absorb the long wavelength radiation reflected back from the ground trapping heat and causing an increase in temperature. Common greenhouse gases are water vapour, carbon dioxide and methane.

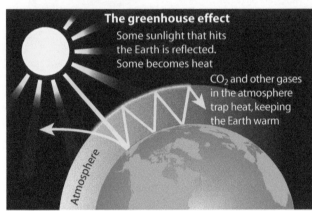

The greenhouse effect

Some sunlight that hits the Earth is reflected. Some becomes heat

CO_2 and other gases in the atmosphere trap heat, keeping the Earth warm

Atmosphere

The increase in carbon dioxide levels in the last century or so correlates with the increased use of fossil fuels by humans.

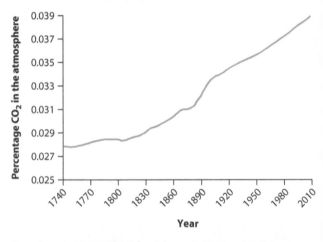

Based on peer-reviewed evidence, many scientists believe that increasing these human activities will lead to global climate change.

Predicting the impact of changes on global climate change is not easy because of the many different contributing factors involved. This can lead to simplified models and speculation often presented in the media that may not be based on all of the evidence. Some people misrepresent the evidence to suit their own needs; this can lead to bias.

Global climate change

Increasing average global temperature is a major cause of climate change. Potential effects of climate change include:

| Rising sea levels, which may cause flooding and coastal erosion | Changes in the food producing capacity of some regions | More frequent and severe storms | Changes to the amount, timing and distribution of rainfall | Changes to the distribution of wildlife species | Temperature and water stress for humans and wildlife |

(ws) Reducing the carbon footprint

The carbon footprint is a measure of the total amount of carbon dioxide (and other greenhouse gases) emitted over the life cycle of a product, service or event.

Problems of trying to reduce the carbon footprint include…

- disagreement over the causes and consequences of global climate change
- lack of public information and education
- lifestyle changes, e.g. greater use of cars / aeroplanes
- economic considerations, i.e. the financial costs of reducing the carbon footprint
- incomplete international co-operation.

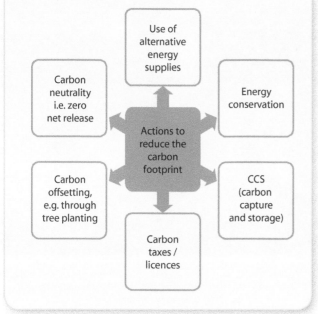

SUMMARY

- **Greenhouse gases maintain the temperature on Earth.**
- **Human activities increase the amount of greenhouse gases, leading to climate change and global warming.**

QUESTIONS

QUICK TEST

1. Name two greenhouse gases.
2. Suggest two potential effects of global climate change.
3. What is a carbon footprint?

EXAM PRACTICE

1. Explain how human activity can contribute to the emission of the greenhouse gases methane and carbon dioxide. **[3 marks]**

Atmospheric pollution

Pollutants from fuels

The combustion of fossil fuels is a major source of atmospheric pollutants. Most fuels contain carbon and often sulfur is present as an impurity. Many different gases are released into the atmosphere when a fuel is burned.

Solid particles and unburned hydrocarbons can also be released forming particulates in the air.

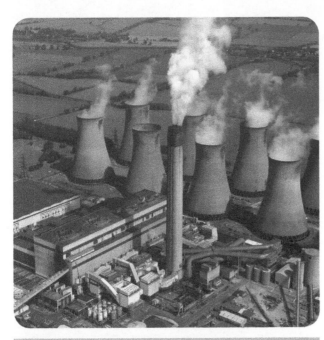

Sulfur dioxide is produced by the oxidation of sulfur present in fuels – often from coal-burning power stations

carbon monoxide

carbon dioxide

Gases produced by burning fuels

oxides of nitrogen

water vapour

sulfur dioxide

Carbon monoxide and soot (carbon) are produced by incomplete combustion of fuels

Oxides of nitrogen are formed from the reaction between nitrogen and oxygen from the air – often from the high temperatures and sparks in the engines of motor vehicles

Properties and effects of atmospheric pollutants

Carbon monoxide is a colourless, odourless toxic gas and so it is difficult to detect. It combines with haemoglobin in the blood, which reduces the oxygen-carrying capacity of blood.

Sulfur dioxide and **oxides of nitrogen** cause respiratory problems in humans and can form acid rain in the atmosphere. Acid rain damages plants and buildings.

Particulates in the atmosphere can cause global dimming, which reduces the amount of sunlight that reaches the Earth's surface. Breathing in particulates can also damage lungs, which can cause health problems.

SUMMARY

- Combustion of fossil fuels is a major source of atmospheric pollutants.
- Gases produced by burning fuels include carbon monoxide, carbon dioxide, sulfur dioxide, water vapour and oxides of nitrogen.

QUESTIONS

QUICK TEST

1. Name three gases produced by burning fuels.

2. How is sulfur dioxide formed?

EXAM PRACTICE

1. Cars over three years old are required to have an annual test to check the roadworthiness of the vehicle.

   As part of this test the amounts of the following pollutants in exhaust gases are measured:
   Carbon monoxide
   Hydrocarbons
   Oxides of nitrogen
   Particulates

   a) Suggest why carbon monoxide may be present in exhaust gases. **[1 mark]**

   b) Suggest why older cars may have higher levels of hydrocarbons in their exhaust gases than new cars. **[1 mark]**

   c) Why is it harmful to breathe in oxides of nitrogen? **[1 mark]**

   d) Why are particulates in the atmosphere harmful? **[2 marks]**

Using the Earth's resources and obtaining potable water

Earth's resources

We use the Earth's **resources** to provide us with warmth, shelter, food and transport. These needs are met from natural resources which, supplemented by agriculture, provide food, timber, clothing and fuels. Resources from the earth, atmosphere and oceans are processed to provide energy and materials.

Example

Chemistry plays a role in providing sustainable development. This means that the needs of the current generation are met without compromising the potential of future generations to meet their own needs. For example…

Industrial chemical processes can be used to make new materials, reducing the demand for natural resources

Chemistry plays an important role in improving agricultural processes, e.g. by developing fertilisers

Drinking water

Water that is safe to drink is called potable water. It is not pure in the chemical sense (see page 60) because it contains dissolved minerals and ions.

Water of appropriate quality is essential for life. This means that it contains sufficiently low levels of dissolved salts and microbes.

In the UK, most potable water comes from rainwater. To turn rainwater into potable water, water companies carry out a number of processes.

1 An appropriate source of fresh water is selected

2 It is then passed through filter beds to remove any solid impurities

3 Finally it is sterilised (suitable sterilising agents include chlorine, ozone and ultraviolet light) to kill microbes, making it safe to drink

Waste water

Urban lifestyles and industrial processes generate large quantities of waste water, which requires treatment before being released into the environment. Sewage and agricultural waste water require removal of organic matter and harmful microbes. Industrial waste water may require removal of organic matter and harmful chemicals.

Sewage treatment includes…

- screening and grit removal
- sedimentation to produce sewage sludge and effluent
- anaerobic digestion of sewage sludge
- aerobic biological treatment of effluent.

When supplies of fresh water are limited, removal of salt (desalination) of salty water / seawater can be used.

This is done in two ways:

- By distillation.

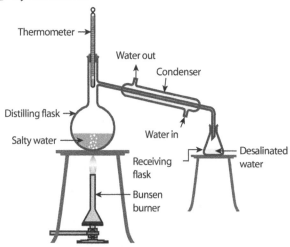

- By processes that use membranes, such as reverse osmosis.

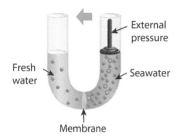

However, these processes require large amounts of energy.

SUMMARY

- Humans use the Earth's resources to supply them with their needs.
- Water is an important resource. Water goes through processes to become safe to drink (potable water).
- Modern life creates a lot of waste water, which must be treated before being released into the environment.

QUESTIONS

QUICK TEST

1. Explain the meaning of the term 'sustainable development'.

2. What is the difference between pure water and potable water?

3. Outline the main stages in the treatment of sewage.

EXAM PRACTICE

1. The diagram below shows how water can be separated from sea water.

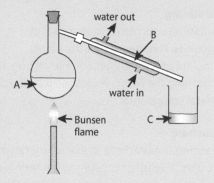

   a) Name this method of desalination. **[1 mark]**

   b) Identify the labels A, B and C. **[3 marks]**

   c) Name another method of removing salt from sea water. **[1 mark]**

   d) Suggest why both of these methods are expensive to carry out. **[1 mark]**

HT Alternative methods of extracting metals

Extracting copper

Copper is an important metal with lots of uses. It is used in electrical wiring because it is an excellent conductor of electricity. It is also used in water pipes because it conducts heat and does not corrode or react with the water.

copper ore

The Earth's resources of metal ores are limited and copper ores are becoming scarce.

New ways of extracting copper from low-grade ores include…

● **phytomining**
Phytomining uses plants to absorb metal compounds. This means that the plants accumulate metal within them. Harvesting and then burning the plants leaves ash that is rich in the metal compounds.

● **bioleaching**
Bioleaching uses bacteria to extract metals from low-grade ores. A solution containing bacteria is mixed with a low-grade ore. The bacteria release the metals into solution (known as a leachate) where they can be easily extracted.

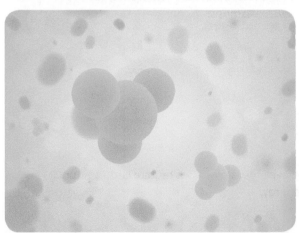

These new extraction methods reduce the impact on the Earth of mining, moving and disposing of large amounts of rock.

Processing metal compounds

The metal compounds from phytomining and bioleaching are processed to obtain the metal. Copper can be obtained from solutions of copper compounds by…

- **displacement** using scrap iron – iron is more reactive than copper, so placing iron into a solution of copper will result in copper metal being displaced.

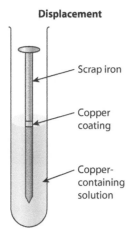

Displacement

Scrap iron

Copper coating

Copper-containing solution

- **electrolysis**.

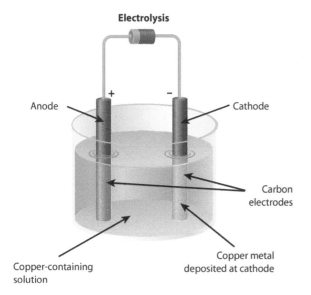

Electrolysis

Anode

+

−

Cathode

Carbon electrodes

Copper-containing solution

Copper metal deposited at cathode

QUESTIONS

QUICK TEST

1. Suggest one use of copper.

2. Name two ways of extracting copper from low-grade ones.

3. What is bioleaching?

4. Give one way that metal can be extracted from a metal containing solution.

EXAM PRACTICE

1. **a)** Explain why alternative methods of extracting copper such as bioleaching are being developed.   **[2 marks]**

    b) A solution containing copper ions is obtained from a bioleaching plant.

    Describe two ways that copper metal can be obtained from this solution.   **[2 marks]**

    c) Other than mining and bioleaching, name another method of obtaining copper from ores.   **[1 mark]**

Life-cycle assessment and recycling

ⓦⓢ Life-cycle assessments

Life-cycle assessments (LCAs) are carried out to evaluate the environmental impact of products in each of the following stages.

| Extracting and processing raw materials | Manufacturing and packaging | Disposal at the end of useful life | Transport and distribution at each of the previous stages |

The following steps are considered when carrying out an LCA.

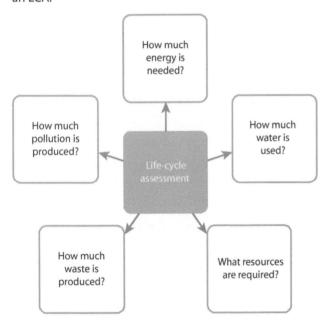

Many of these values are relatively easy to quantify. However, some values, such as the amount of pollution, are often difficult to measure and so value judgements have to be made. This means that carrying out an LCA is not a purely objective process.

It is not always easy to obtain accurate figures. This means that selective or abbreviated LCAs, which are open to bias or misuse, can be devised to evaluate a product, to reinforce predetermined conclusions or to support claims for advertising purposes.

For example, look at the LCA below.

Example of an LCA for the use of plastic (polythene) and paper shopping bags		
	Amount per 1000 bags over the whole LCA	
	Paper	Plastic (polythene)
Energy use (MJ)	2590	713
Fossil fuel use (kg)	28	13
Solid waste (kg)	34	6
Greenhouse gas emissions (kg CO_2)	72	36
Freshwater use (litres)	3387	198

This LCA provides evidence supporting the argument that using polythene bags is better for the environment than paper bags!

Ways of reducing the use of resources

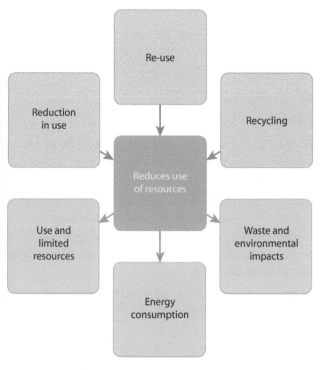

- Re-use
- Reduction in use
- Recycling
- Reduces use of resources
- Use and limited resources
- Waste and environmental impacts
- Energy consumption

Many materials such as glass, metals, building materials, plastics and clay ceramics are produced from limited raw materials. Most of the energy used in their production comes from limited resources, such as fossil fuels. Obtaining raw materials from the earth by quarrying and mining has a detrimental environmental impact.

Some products, like glass, can be **reused** (e.g. washing and then using again for the same purpose). Recycled glass is crushed, melted and remade into glass products.

Other products cannot be reused and so they are **recycled** for a different use.

Metals are recycled by sorting them, followed by melting them and recasting / reforming them into different products. The amount of separation required for recycling depends on the material (e.g. whether they are magnetic or not) and the properties required of the final product. For example, some scrap steel can be added to iron from a blast furnace to reduce the amount of iron that needs to be extracted from iron ore.

SUMMARY

- LCAs are carried out to evaluate the environmental impact of products at each stage in their production and distribution.
- Recycling is important in order to reduce the use of resources, and therefore reduce energy use, waste and environmental impacts.

QUESTIONS

QUICK TEST

1. What does a life-cycle assessment measure?
2. State two factors that a life-cycle assessment tries to evaluate.
3. What is a blast furnace?
4. Suggest one way in which we can reduce the use of resources.

EXAM PRACTICE

1. A Life Cycle Assessment (LCA) comparing paper and plastic shopping bags was carried out by an independent scientist.

   The report stated:

   'Production of a paper bag produces 2.5 times the amount of carbon dioxide than a plastic bag.'

   A journalist quoted this fact in an article and wrote:

   'Based on this fact, on environmental grounds we should be using plastic bags.'

   Evaluate this statement using the information provided and your own knowledge. **[3 marks]**

Using materials

Corrosion and its prevention

Corrosion is the destruction of materials (e.g. metals) by chemical reactions in the environment. Rusting is a common form of corrosion. It involves iron reacting with oxygen and water.

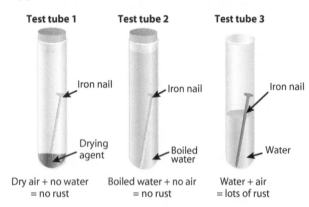

Test tube 1	Test tube 2	Test tube 3
Iron nail / Drying agent	Iron nail / Boiled water	Iron nail / Water
Dry air + no water = no rust	Boiled water + no air = no rust	Water + air = lots of rust

Rusting can be prevented by applying a coating that acts as a barrier, such as…

- greasing
- painting
- electroplating.

Aluminium is a reactive metal that naturally forms an oxide coating which prevents the metal from further corrosion.

Some coatings are reactive and may contain corrosion inhibitors or a more reactive metal. For example, zinc is often used to galvanise iron. If the coating is then scratched, the zinc reacts instead of the iron. This is called **sacrificial protection**.

Magnesium blocks can be attached to a ship's hull to provide sacrificial protection.

Alloys as useful materials

Many of the metals we use every day are alloys. Pure metals such as gold, copper, iron and aluminium are too soft for many uses. Alloys are made by mixing metals together. This makes them harder and more suitable for everyday use.

100% pure **gold** (known as 24-carat gold) is often mixed with other metals. The proportion of gold in the alloy gives the carat rating (e.g. 18 carat contains 75% gold).	
Bronze (an alloy of copper and tin) is used to make statues and other decorative objects.	
Brass is an alloy of copper and zinc that is used to make water taps and door fittings.	
Steel is an alloy of iron, carbon and other metals such as chromium or nickel.	
High-carbon steel is strong but brittle – it is used to make cutting tools.	
Low-carbon steel is softer, so it can be more easily shaped. It is used to make car body panels.	
Steels containing nickel and chromium are known as **stainless steels** because of their resistance to corrosion. They are hard and are used to make cutlery and railway tracks.	
Aluminium alloys are strong and low in density. They are used in aerospace manufacturing, for example to make the bodies of aeroplanes.	

Ceramics

Everyday glass is soda-lime glass.

> sand + sodium carbonate + limestone → soda-lime glass

Borosilicate glass, made from sand and boron trioxide, melts at a higher temperature than soda-lime glass.

Clay ceramics, including pottery and bricks, are made by shaping wet clay and then heating it in a furnace.

Polymers

The properties of polymers depend on the monomer used to make the polymer and the conditions under which they are made. Low-density and high-density poly(ethene) are made from ethene, but using different catalysts and reaction conditions.

Thermosoftening polymers consist of individual polymer chains all tangled together which melt when heated. Thermosetting polymers consist of cross-links between the polymer chains and so they have a more rigid structure which means that they do not melt easily when they are heated.

Thermosoftening polymer (no cross-links)

Thermosetting polymer

Cross-links

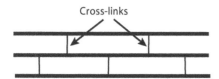

Composites

Most composite materials, such as reinforced concrete, wood and fibreglass, consist of two materials: a matrix or binder and a reinforcement substance.

Reinforced concrete column

Some advanced composites are made from carbon fibres or carbon nanotubes instead of glass fibres.

SUMMARY

● Chemicals in the environment can react to produce corrosion of materials, e.g. rusting.
● Corrosion can be prevented by applying coatings.
● Alloys are made by mixing metals, which makes them harder.
● Polymers have different properties depending on the monomers used and the conditions under which they're made.

QUESTIONS

QUICK TEST

1. Rusting is when iron reacts with which two elements?

2. What property does stainless steel have compared with ordinary steel?

3. What is an alloy?

4. What do sand, limestone and sodium carbonate produce when reacted together?

EXAM PRACTICE

1. Suggest a suitable coating to prevent rusting for each of the following items.

   [3 marks]

   a) Bicycle chain
   b) A bridge
   c) Iron nail

2. For a named alloy, state its composition and how the properties of the alloy relate to its use. **[3 marks]**

The Haber process and the use of NPK fertilisers

The Haber process

Ammonia (NH_3) is produced by the Haber process. The ammonia is used to make nitrogen-based fertilisers.

Ammonia is made by combining nitrogen (obtained by fractional distillation of liquid air) and hydrogen (from natural gas).

$$\text{nitrogen} + \text{hydrogen} \rightleftharpoons \text{ammonia}$$
$$N_{2(g)} + 3H_{2(g)} \rightleftharpoons 2NH_{3(g)}$$

The purified gases are passed over an iron catalyst. The reaction is carried out at 450 °C and 200 atmospheres pressure. The reaction is reversible, so some of the ammonia produced will break down into nitrogen and hydrogen.

Upon cooling the ammonia liquefies and is removed. The unreacted nitrogen and hydrogen are recycled back into the reaction chamber.

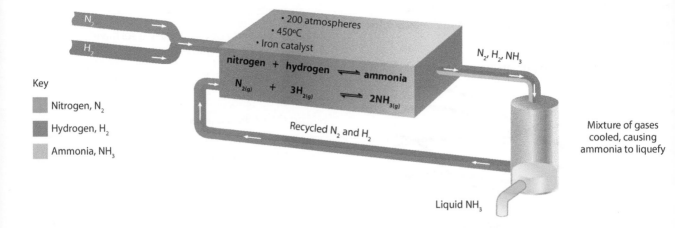

Key
- Nitrogen, N_2
- Hydrogen, H_2
- Ammonia, NH_3

Production and uses of NPK fertilisers

NPK fertilisers contain the elements nitrogen (chemical symbol **N**), phosphorus (**P**) and potassium (**K**). NPK fertilisers supply three important chemical elements that plants need and so the addition of NPK fertilisers improves agricultural productivity.

NPK fertilisers are formulations (see page 60) of various salts containing appropriate percentages of N, P and K. Production of fertilisers can be achieved by using a variety of raw materials in several integrated processes.

Ammonia can be used to manufacture nitric acid (HNO_3) and ammonium salts, such as ammonium sulfate, $(NH_4)_2SO_4$.

Potassium chloride, potassium sulfate and phosphate rock are obtained by mining. Phosphate rock requires further processing as it is insoluble and so cannot be used directly as a fertiliser.

Phosphate rock is treated with nitric acid to make phosphoric acid (H_3PO_4) and calcium nitrate, $Ca(NO_3)_2$. Phosphoric acid is neutralised with ammonia to make ammonium phosphate.

$$H_3PO_4 + 3NH_3 \rightarrow (NH_4)_3PO_4$$

When phosphate rock is reacted with sulfuric acid, 'single super phosphate' is produced. This is a mixture of calcium phosphate and calcium sulfate.

If phosphate rock is reacted with phosphoric acid then 'triple superphosphate' calcium phosphate is produced.

HT Explaining the conditions used in the Haber process

The reaction to produce ammonia is exothermic:

$$N_{2(g)} + 3H_{2(g)} \rightleftharpoons 2NH_{3(g)} \qquad \Delta H = -92 \text{ kJmol}^{-1}$$

High temperatures favour endothermic reactions (see page 51) so if the temperature is too high then the yield (amount produced) of ammonia will be low. If the temperature is too low then the yield of ammonia will be higher but the rate will be too slow. 450°C is a compromise temperature between rate and yield.

High pressure favours the reaction that produces fewer molecules of gas (in this case, high pressure favours the forward reaction, i.e. the production of ammonia). If the pressure is increased, the yield of ammonia and the rate of production will both increase but so will the cost of production. The increased revenue from the sale of the extra ammonia does not compensate for the additional costs. 200 atmospheres pressure is a compromise between yield, rate and cost.

Different Haber process plants operate at slightly different conditions based upon…

● the availability and cost of raw materials and energy

● the demand for ammonia.

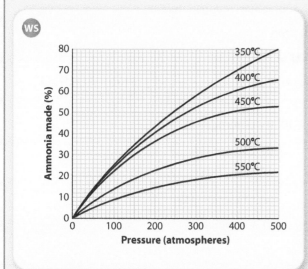

SUMMARY

● The Haber process produces ammonia.

● Ammonia is used to make nitrogen-based fertilisers.

● NPK fertilisers contain nitrogen, phosphorus and potassium.

QUESTIONS

QUICK TEST

1. Which raw materials are needed to make ammonia?

2. At what temperature is the Haber process carried out?

EXAM PRACTICE

1. Ammonia, NH_3, can be used to make compounds for use in NPK fertilisers. One compound commonly found in NPK fertilisers is ammonium sulfate, $(NH_4)_2SO_4$.

   a) What three elements are found in NPK fertilisers? **[1 mark]**

   b) Write a balanced symbol equation for the reaction between ammonia and sulfuric acid, H_2SO_4, forming ammonium sulfate.

   State symbols are not required. **[1 mark]**

HT 2. The Haber process typically operates at a temperature of 450°C and 200 atmospheres.

   State and explain the effect on the rate of production and yield of ammonia of changing the temperature to 300°C and a pressure of 250 atmospheres.

   The equation for the reaction taking place in a Haber plant is:

   $$N_{2\,(g)} + 3H_{2\,(g)} \rightleftharpoons 2NH_{3\,(g)} \quad \Delta H = -92 \text{ kJ/mol}$$

   [6 marks]

Answers

Page 5
QUICK TEST

1. Atoms

2. A compound

3. Crystallisation

EXAM PRACTICE

1. a) $2Mg_{(s)} + O_{2\,(g)} \rightarrow 2MgO_{(s)}$

   Correctly balanced equation [1] Correct state symbols [1]

   b) Two or more elements (or compounds) that are together but not chemically combined. [1]

   c) Add water to the mixture and stir. [1] Filter to remove the insoluble magnesium oxide. [1] Heat/leave the magnesium chloride solution somewhere warm to allow the water to evaporate. [1]

Page 7
QUICK TEST

1. +1 / positive

2. 12 protons, 12 electrons and 12 neutrons

3. Isotopes

EXAM PRACTICE

1. In the plum pudding model, the electrons are embedded throughout a ball of positive charge. [1] In the nuclear model, there is a central nucleus containing the positive charge [1] and the electrons are located in energy levels/shells. [1]

Page 9
QUICK TEST

1. 2, 8, 8, 1

2. He also predicted that there were more elements to be discovered. He predicted their properties and left appropriate places in the periodic table based on his predictions.

EXAM PRACTICE

1. a) 2,8,2 [1]

   b) After. [1] Element X has 12 electrons (and therefore 12 protons). The periodic table is arranged by increasing atomic number and 14 is greater than 12. [1]

Page 11
QUICK TEST

1. **Examples:** Airships, balloons, light bulbs, lasers, advertising signs

2. Potassium hydroxide and hydrogen

3. The reactivity decreases.

EXAM PRACTICE

1. a) **Three from:** Sodium floats; Moves about the surface; Melts; Fizzes/effervescence [3]

   b) $2Na_{(s)} + 2H_2O_{(l)} \rightarrow 2NaOH_{(aq)} + H_{2\,(g)}$
   Correct formulae [1] Correct state symbols [1]

2. a) Bromine is formed [1] which is brown. [1]

   b) Chlorine is more reactive than bromine [1] and so will take the place of/displace bromine in a compound. [1]

Page 13
QUICK TEST

1. **Two from:** Transition metals have higher melting points (except for mercury); are more dense; are less reactive with water and oxygen.

2. Potassium manganate(VII)

3. Iron

EXAM PRACTICE

1. a) Both conduct electricity or heat or both are malleable/ductile [1]

   b) i) Chromium [1]

   ii) Sodium [1]

   iii) Sodium [1]

   iv) Chromium [1]

   v) Chromium [1]

QUICK TEST

1. Covalent

2. Two

3. A free-moving electron

4. The electrostatic force of attraction between two oppositely charged ions

EXAM PRACTICE

1. a) Potassium atoms lose their outer shell electron and sulfur atoms gain two electrons [1] forming K^+ ions that have the electronic configuration 2,8,8 [1] and S^{2-} ions that have the electronic configuration 2,8,8 [1]

   b) K_2S [1]

2. a)

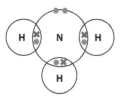

   3 atoms of hydrogen each with two electrons. [1]
   Pair of electrons in the outer shell of nitrogen. [1]

   b) Atoms of neon have full outer shells of electrons [1] and so don't need to gain, lose or share any electrons. [1]

3. A lattice of metal cations [1] surrounded by delocalised electrons. [1]

Page 17
QUICK TEST

1. **Any metal bonded to a non-metal, e.g.** sodium chloride

2. Electrostatic forces/strong forces between cations and anions

3. Giant covalent/macromolecular, simple molecular, polymer

EXAM PRACTICE

1. a) XCl_2 [1] There are twice as many chloride ions as there are X ions. [1]

   b) **One from:** Only a few ions are shown; The ions are not connected by 'lines'/chemical bonds are not solid objects; The ions are not held apart; Diagram doesn't show relative size of particles. [1]

2. a) Covalent [1]

   b) Simple molecular [1]

   c) Giant covalent [1]

Page 19
QUICK TEST

1. solid, liquid, gas

2. The intermolecular forces/forces between the molecules

EXAM PRACTICE

1. a) −40°C [1]

   b) Liquid [1]

   c) Condensing/condensation [1]

2. In a solid the ions are in a fixed position/unable to move. [1] In the liquid state the ions are free to move [1] and carry the charge. [1]

Page 21
QUICK TEST

1. Because there are strong forces of attraction between the metal cations and delocalised electrons.

2. **Two from:** Drug delivery into the body; Lubricants; Reinforcing materials

EXAM PRACTICE

1. a) A mixture of two or more metals, or a mixture of a metal and a non-metal. [1]

   b) The layers of atoms are not able to slide over each other easily [1] because the different atoms have different sizes [1] and so the layers aren't regular.

2. a) Layers of carbon atoms [1] that each form three covalent bonds to other carbon atoms [1] There are weak intermolecular forces between the layers. [1]

   b) Each carbon atom only forms three bonds/has a spare/delocalised electron [1] which can move and carry the charge. [1]

Page 23
QUICK TEST

1. Less than 100 nm (1×10^{-7} m)

2. It increases by a factor of 10.

3. Better skin coverage/more effective protection from the Sun's UV rays.

Answers

EXAM PRACTICE

1. **a) Two from:** Development of new catalysts for fuel cell materials; Controlled drug delivery; Cosmetics and sun creams; Synthetic skin; Electronics; Deodorants **[2]**

   b) There are concerns over potential harmful effects when nanoparticles are released into the environment **[1]** and over potential cell damage in the body. **[1]**

Page 25
QUICK TEST

1. The same

2. The sum of the atomic masses of the atoms in a formula

3. Lower

4. 58

EXAM PRACTICE

1. **a)** 84 **[1]**

   b) $(10 - 4.76) = 5.24$ g **[1]**

   c) Thermal decomposition or endothermic **[1]**

   d) It will increase **[1]** because oxygen is being added to the metal. **[1]**

Page 27
QUICK TEST

1. Li_2O

2. P_4O_{10}

EXAM PRACTICE

1. **Ratio method:**

	4Al		3O$_2$	2Al$_2$O$_3$
Number of moles reacting	4		3	2
Relative formula mass	27		32	102
Mass reacting/formed (g)	108	÷25	96	204 ÷25
Reacting mass	4.32			8.16

Correct answer [3]

Moles method:

Number of moles of aluminium reacting

$= \dfrac{4.32}{27} = 0.16$ **[1]**

Due to 4:2 ratio the number of moles of aluminium oxide formed = 0.08 **[1]**
Therefore the mass of aluminium oxide formed $= 0.08 \times 102 = 8.16$ g **[1]**

Page 29
QUICK TEST

1. A measure of the amount of starting materials that end up as useful products.

2. **One from:** The reaction may not go to completion because it is reversible; Some of the products may be lost when separated from the reaction mixture; Some of the reactants may react in ways that are different to the expected reaction.

EXAM PRACTICE

1. **a)** $\dfrac{2 \times [1 + 14 + (3 \times 16)]}{3 \times [14 + (2 \times 16)] + [2 \times 1 + 16]} \times 100$ **[1]**

   $= \dfrac{126}{156} \times 100 = 80.8$ % **[1]**

   b) $\dfrac{12.1}{17.5} \times 100 = 69.1\%$ **[2]**

   c) The reaction has a higher/100% atom economy **[1]** meaning that there is no waste formed/which reduces the demand for raw materials, **[1]** which is more sustainable. **[1]**

Page 31
QUICK TEST

1. 0.84 g

2. 0.2 mol/dm^3

EXAM PRACTICE

1. **a)** $\dfrac{0.10 \times 18.60}{1000}$ **[1]** $= 1.86 \times 10^{-3}$ **[1]**

   b) 1:2 mole ratio **[1]** therefore the number of moles of sulfuric acid

   $= \dfrac{1.86 \times 10^{-3}}{2} = 9.3 \times 10^{-4}$ **[1]**

   c) $\dfrac{9.3 \times 10^{-4} \times 1000}{25}$ **[1]**

   $= 0.0372$ mol/dm^3 **[1]**

   d) 0.0372×98 **[1]** $= 3.65$ g/dm^3 **[1]**

   Therefore, the labelled concentration is incorrect/inaccurate or the actual concentration is not the same as the labelled concentration. **[1]**

Page 33
QUICK TEST

1. 0.6375

2. $Si + 2Cl_2 \rightarrow SiCl_4$

EXAM PRACTICE

1. Number of moles of methane gas $= \frac{3}{24} = 0.125$ **[1]**

   1:2 mole ratio therefore number of moles of oxygen required $= 2 \times 0.125 = 0.25$ **[1]**

   Volume $= 0.25 \times 24 = 6$ dm^3 **[1]**

   This question can be done by ratios. 1 dm^3 of CH_4 will react with 2 dm^3 of O_2 therefore 3 dm^3 of CH_4 reacts with 6 dm^3 of O_2. **Correct answer scores [3]**

2. Moles of lithium reacting $= \frac{4.2}{7} = 0.6$ **[1]**

   6:1 mole ratio therefore the number of moles of nitrogen required $= 0.1$ **[1]**

   Volume of nitrogen $= 0.1 \times 24\,000 = 2\,400$ cm^3 **[1]**
   [Also accept 2.4 dm³] [1]

Page 35
QUICK TEST

1. **Two from**: Zinc; Iron; Copper

2. Magnesium

3. aluminium + oxygen $\rightarrow$ aluminium oxide

4. Sodium (because it is more reactive)

EXAM PRACTICE

1. a) Magnesium is more reactive than copper. **[1]**

   b) Copper(II) oxide **[1]** as it loses oxygen. **[1]**
      [Allow copper/Cu^+ **[1]** as it gains electrons **[1]]**

Page 37
QUICK TEST

1. A method used to separate a soluble solid from its solution when you want to collect the solid.

2. Fe

3. aluminium + sulfuric acid $\rightarrow$ aluminium sulfate + hydrogen.

4. Zinc nitrate

5. Calcium sulfate

EXAM PRACTICE

1. a) $MgO_{(s)} + 2HCl_{(aq)} \rightarrow MgCl_{2\,(aq)} + H_2O_{(l)}$
      Correct formulae [1] Correctly balanced [1]

   b) **Five from:** Using a measuring cylinder, transfer a known volume of hydrochloric acid to a beaker; Warm the acid; Add a spatula of magnesium oxide powder and stir until it dissolves; Repeat until no more magnesium oxide dissolves; Filter the mixture to remove excess magnesium oxide; Heat the filtrate/leave it somewhere warm to remove the water/allow crystallisation to occur. **[5]**

Page 39
QUICK TEST

1. A liquid or solution containing ions that is broken down during electrolysis

2. H^+

3. Alkaline/alkali

4. A strong acid fully ionises/dissociates in solution. A weak acid only partially ionises/dissociates in solution.

5. Cations

EXAM PRACTICE

1. a) Pink **[1]** to colourless **[1]**

   b) The hydroxide ion/OH^- **[1]**

   c) $H^+_{(aq)} + OH^-_{(aq)} \rightarrow H_2O_{(l)}$ **[1]**

   d) Strong acids fully/completely **[1]** dissociate/ionise **[1]** in solution.

Page 41
QUICK TEST

1. Copper will be formed at the cathode; chlorine will be formed at the anode.

2. Hydrogen/H^+ ions and hydroxide/OH^- ions.

EXAM PRACTICE

1. Product at cathode = copper **[1]**
   Product at anode = oxygen **[1]**

Page 43
QUICK TEST

1. Exothermic – **One from:** Combustion; Neutralisation; Oxidation; Precipitation; Displacement
   Endothermic – **One from:** Thermal Decomposition; Citric acid; Sodium hydrogen carbonate

2. Given out

Answers

EXAM PRACTICE

1.

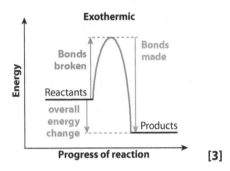

Exothermic

(Graph: Energy vs Progress of reaction; Reactants, Bonds broken, Bonds made, Products, overall energy change) **[3]**

2. Bonds broken:

H–H   436

Cl–Cl   239

Total = 675 **[1]**

Bonds formed:

2 H—Cl = 2 × 427 = 854 **[1]**

Energy change = 675 – 854 = –179 kJ/mol **[1]**

Page 45
QUICK TEST

1. A chemical cell where the source of fuel is supplied externally/continuously

2. A battery is a combination of chemical cells connected in series.

3. $H_{2(g)} \rightarrow 2H^+_{(aq)} + 2e^-$

EXAM PRACTICE

1. a) They are constantly supplied with fuel. **[1]**

b) Water/H_2O **[1]**

c) **One from:** Hydrogen is difficult to obtain; It is dangerous; It is difficult to store. **[1]**

d) $4H^+_{(aq)} + O_{2\,(g)} + 4e^- \rightarrow 2H_2O_{(g)}$ **[1]**

Page 47
QUICK TEST

1. 3 g/s

2. **Two from:** The concentrations of the reactants in solution; The pressure of reacting gases; The surface area of any solid reactants; Temperature; Presence of a catalyst

EXAM PRACTICE

1. a) $\frac{90}{42}$ **[1]** = 2.1 **[1]** cm³/s

Answer must be to 1 decimal place.

b) It would decrease/halve. **[1]**

c) **One from:** Increasing the surface area of the magnesium/using magnesium powder; Increasing the temperature; Adding a catalyst **[1]**

Page 49
QUICK TEST

1. The minimum amount of energy that the particles must have when they collide in order to react.

2. There are more particles in the same volume of liquid and so there are more chances of reactant particles colliding.

3. The idea that, for a chemical reaction to occur, the reacting particles must collide with sufficient energy.

4. They provide an alternative pathway of lower activation energy.

EXAM PRACTICE

1. a) The particles/molecules have more kinetic energy/move more quickly **[1]** meaning there will be more collisions per unit of time. **[1]** At a higher temperature more of the particles will possess energy equal to or greater than the activation energy for the reaction meaning there will be more successful collisions. **[1]**

b) It provides an alternative reaction pathway **[1]** of lower activation energy. **[1]**

c) At higher pressure there will be more molecules of gas per unit of volume **[1]** meaning more collisions per unit of time. **[1]**

Page 51
QUICK TEST

1. Endothermic

2. Equilibrium moves to the right-hand side / favours the forward reaction.

EXAM PRACTICE

1. $N_2 + 3H_2 \rightleftharpoons 2NH_3$
Correct balanced equation [1] Use of $\rightleftharpoons$ symbol [1]

b) The rate **[1]** of the forward reaction is equal to the rate of the reverse reaction. **[1]**

2. a) Increasing pressure favours the reaction/moves the equilibrium position to the side with the fewer moles of gas. **[1]** Therefore, the equilibrium position will move to the left-hand side/the reverse reaction will be favoured. **[1]**

b) Increasing the temperature favours the endothermic reaction. **[1]** In this case the forward reaction is endothermic (indicated by a positive ΔH value) and so increasing the temperature will move the position of equilibrium to the right-hand side/the forward reaction will be favoured. **[1]**

Page 53
QUICK TEST

1. Dead biomass

2. C_nH_{2n+2}

3. C_2H_6

EXAM PRACTICE

1. a) To separate it into useful fractions/components **[1]**

b) The crude oil is heated until it boils/vaporises. **[1]** It then enters the fractionating column where there is a temperature gradient. **[1]** Molecules condense at different levels/positions in the fractionating column according to their boiling point. **[1]**

Page 55
QUICK TEST

1.

2. A process used to break up large hydrocarbon molecules into smaller, more useful ones.

EXAM PRACTICE

1. a) $C_8H_{18} \rightarrow C_4H_{10} + 2C_2H_4$
Correct formulae [1] Balanced equation [1]

b) The alkene will decolourise **[1]** bromine water. **[1]**

Page 57
QUICK TEST

1. CH_3CH_2OH

2. Ethanol

3.

4. Ethyl ethanoate and water

EXAM PRACTICE

1. a) Methanol **[1]**

b) Hydrogen **[1]**

c) A carboxylic acid/methanoic acid **[1]**

2. a) An alcohol **[1]**

b) The acid partially/incompletely **[1]** ionises/dissociates **[1]** in solution.

Page 59
QUICK TEST

1. Plastic shopping bags/water bottles

2. Two from: DNA; Proteins; Starch; Cellulose

EXAM PRACTICE

1. a) A carbon-carbon double bond **[1]**

b) Condensation polymerisation **[1]**

Page 61
QUICK TEST

1. To separate mixtures of dyes

2. A mixture that has been designed as a useful product

3. Two from: Fuels; Cleaning materials; Paints; Medicines; Foods; Fertilisers

4. 0.7

EXAM PRACTICE

1. a) Green and yellow **[1]**

b) By mixing **[1]** purple and yellow **[1]** inks.

Page 63
QUICK TEST

1. Oxygen

2. Add a lighted splint and there will be a squeaky pop.

3. Limewater

4. It turns milky/cloudy.

EXAM PRACTICE

1. Test the gas with damp **[1]** Litmus paper. **[1]** It will be bleached if chlorine is present. **[1]**

Answers

Page 65
QUICK TEST

1. Lilac

2. Sodium

3. Aluminium hydroxide

4. $Fe^{2+}_{(aq)} + 2OH^-_{(aq)} \rightarrow Fe(OH)_{2(s)}$

EXAM PRACTICE

1. a) A yellow [1] flame will be produced.

   b) The flame colour will be masked [1] due to the presence of both sodium and potassium metals/ions producing a flame colour. [1]

2. Dissolve the sample in (distilled/deionised) water. [1] Add sodium hydroxide solution. [1] A white precipitate will form. [1] If the precipitate dissolves when additional sodium hydroxide is added then the solid was aluminium chloride / If the precipitate doesn't dissolve then the solid was magnesium chloride. [1]

Page 67
QUICK TEST

1. The carbonate (CO_3^{2-}) ion

2. (Pale) yellow

EXAM PRACTICE

1. Dissolve the potassium bromide in deionised/distilled water. [1] Add nitric acid [1] followed by silver nitrate solution. [1] A cream precipitate will be formed.

2. **Two from:** Instrumental methods are: more accurate; more sensitive i.e. can produce clearer results; can be used with very small samples [2]

Page 69
QUICK TEST

1. As a product of photosynthesis

2. **Two from:** Carbon dioxide; Water vapour; Methane; Ammonia; Nitrogen; Sulfur dioxide

3. Approximately 0–3%

4. Carbon dioxide

EXAM PRACTICE

1. Due to photosynthesis. [1] The amount of carbon dioxide has decreased. [1] The amount of oxygen has increased. [1]

2. a) $16 \, cm^3$ [1]

b) **Two from:** Carbon dioxide; Water (vapour); Argon; Any other named noble gas [2]

c) $\frac{164}{820} \times 100 = 20\%$ [1]

Page 71
QUICK TEST

1. **Two from:** Water vapour; Carbon dioxide; Methane

2. **Two from:** Rising sea levels leading to flooding/coastal erosion; More frequent/severe storms; Changes to the amount, timing and distribution of rainfall; Temperature and water stress for humans and wildlife; Changes in the food producing capacity of some regions; Changes to the distribution of wildlife species

3. The total amount of carbon dioxide (and other greenhouse gases) emitted over the full life cycle of a product, service or event.

EXAM PRACTICE

1. **Three from:** Combustion of fossil fuels leads to greater carbon dioxide emissions into the atmosphere; Deforestation reduces the amount of carbon dioxide removed from the atmosphere by photosynthesis; Increased animal farming releases more methane into the atmosphere; Increased use of landfill sites releases more methane into the atmosphere [3]

Page 73

QUICK TEST

1. **Three from:** Carbon dioxide; Carbon monoxide; Water vapour; Sulfur dioxide; Nitrogen oxides

2. From the oxidation/combustion of sulfur in fuels

EXAM PRACTICE

1. a) Incomplete combustion of petrol/diesel/fuel [1]

   b) As the car gets older less oxygen is able to get to burn the fuel e.g. blocked pipes etc. [1]

   c) They can cause respiratory problems. [1]

   d) Global dimming [1] Lung damage [1]

Page 75
QUICK TEST

1. Living such that the needs of the current generation are met without compromising the ability of future generations to meet their own needs.

2. Pure water has no chemicals added to it. Potable water may have other substances in it but it is safe to drink.

3. Screening and grit removal; Sedimentation; Anaerobic digestion of sludge; Aerobic biological treatment of effluent

EXAM PRACTICE

1. **a)** Distillation [1]

   b) A = sea water/salty water [1]

   B = condenser [1]

   C = pure/desalinated water [1]

   c) Reverse osmosis [1]

   d) They require lots of energy. [1]

Page 77
QUICK TEST

1. **One from**: Electrical wiring; Water pipes

2. Phytomining, bioleaching

3. An extraction method that uses bacteria to extract metals from low-grade ores.

4. **One from**: Displacement using scrap iron; Electrolysis

EXAM PRACTICE

1. **a)** Supplies of copper rich ores are diminishing. [1] Bioleaching is a method of obtaining copper from low-grade ores. [1]

   b) By displacement/adding a more reactive metal e.g. (scrap) iron. [1] By electrolysis/passing electricity through the solution. [1]

   c) Phytomining [1]

Page 79
QUICK TEST

1. The environmental impact of a product over the whole of its life

2. **Two from**: How much energy is needed; How much water is used; What resources are required; How much waste is produced; How much pollution is produced

3. An industrial method of extracting iron from iron ore

4. **One from**: Using fewer items that come from the earth; Reusing items; Recycling more of what we use

EXAM PRACTICE

1. **Three from:** The production of carbon dioxide is only one measure in a LCA; Other pollutants may be produced in the production of plastic bags; No reference is made of energy requirements; No reference is made to water usage; Plastic bags may be used many more times than paper bags/plastic

bags have a longer life span than paper bags; The LCA is incomplete; More trees may be grown to produce more paper for bags which absorb CO_2 **[3]**

Page 81
QUICK TEST

1. Oxygen and water

2. It is more resistant to corrosion/rusting.

3. A mixture of metals/A metal mixed with another element

4. Soda-lime glass

EXAM PRACTICE

1. **a)** Grease/oil [1]

   b) Painting [1]

   c) Galvanising/coating with zinc [1]

2. Name e.g. stainless steel [1]
   Composition e.g. iron with chromium/nickel [1]
   Property e.g. used in cutlery as it doesn't rust/corrode [1]
   (See table on page 80 for other alloys.)

Page 83
QUICK TEST

1. Nitrogen and hydrogen

2. 450°C

EXAM PRACTICE

1. **a)** Nitrogen, phosphorus and potassium [1]

   b) $2NH_3 + H_2SO_4 \rightarrow (NH_4)_2SO_4$ [1]

2. **Six from:** Decreasing temperature will increase the yield of ammonia; because decreasing the temperature favours the exothermic reaction which in this case is the forward reaction; the rate of reaction will decrease; because at a lower temperature the particles have less energy and so there will be fewer collisions; and a lower proportion of successful collisions because fewer molecules will possess the activation energy required to react; a higher pressure will increase the rate of reaction; because there will be more collisions because there are more particles per unit volume; and the yield of ammonia will increase; because higher pressure favours the reaction that produces fewer molecules of gas; which in this case is the forward reaction **[6]**

Glossary

Activation energy – The minimum amount of energy that particles must collide with in order to react

Addition polymerisation – The process where many small unsaturated monomers join together to form a polymer

Alcohol – A molecule containing the –OH functional group

Alloy – A mixture of two or more metals, or a mixture of a metal and a non-metal

Anion – A negative ion

Anode – The positive electrode

Atom economy – A measure of the amount of starting materials that end up as useful products

Atom – The smallest part of an element that can enter into a chemical reaction

Atomic number – The number of protons in the nucleus of an atom

Avogadro's constant – 6×10^{23} (the number of particles in one mole)

Battery – A combination of cells connected in series

HT Bioleaching – An extraction method that uses bacteria to extract metals from low-grade ores

Blast furnace – Industrial method of extracting iron from iron ore

Carbon footprint – The total amount of carbon dioxide (and other greenhouse gases) emitted

Carboxylic acid – A molecule containing the –COOH functional group

Catalyst – A substance that changes the rate of a chemical reaction without being used up or chemically changed at the end of the reaction

Cathode – The negative electrode

Cation – A positive ion

Chemical cell – A system containing chemicals that react together to produce electricity

Chromatography – A method of separating mixtures of dyes

Composite material – Two (or more) different materials combined together

Compound – A substance consisting of two or more elements chemically combined together

HT Condensation polymerisation – The process where many small molecules join together forming a polymer and a small molecule such as water

Cracking – A process used to break up large hydrocarbon molecules into smaller, more useful molecules.

Crystallisation – A method used to separate a soluble solid from its solution when you want to collect the solid

Delocalised electrons – Free-moving electrons

Displacement reaction – A reaction in which a more reactive element takes the place of a less reactive element in a compound

Electrolyte – A liquid or solution containing ions that is broken down during electrolysis

Element – A substance that consists of only one type of atom

Empirical formula – The simplest whole number ratio of each kind of atom present in a compound

Endothermic – A reaction in which energy is taken in

Energy level – A region in an atom where electrons are found

Equilibrium – A reversible reaction where the rate of the forward reaction is the same as the rate of the reverse reaction

Exothermic – A reaction in which energy is given out

Fermentation – The process by which yeast converts sugars to alcohol and carbon dioxide through anaerobic respiration

Fertiliser – Any material added to the soil or applied to a plant to improve the supply of minerals and increase crop yield

Formulation – A mixture that has been designed as a useful product

Fossil fuel – Fuel formed in the ground over millions of years from the remains of dead plants and animal

Fractional distillation – A method used to separate mixtures of liquids

Fuel cell – A chemical cell where the source of fuel is supplied externally

Fullerene – A molecule made of carbon atoms arranged as a hollow sphere

Functional group – The group of atoms in a molecule that determines the chemical behaviour of the compound

Galvanise – Protect a metal by coating it with zinc

Haber process – A process used to make ammonia, NH_3

Halogen – One of the five non-metals in group 7 of the periodic table

High tensile strength – Does not break easily when stretched

Hydrocarbon – A molecule containing hydrogen and carbon atoms only

Intermolecular forces – The weak forces of attraction that occur between molecules

Ion – An atom or group of atoms that has gained or lost one or more electrons in order to gain a full outer shell

Isotopes – Atoms of the same element that have the same number of protons but different numbers of neutrons

Life-cycle assessment – An evaluation of the environmental impact of a product over the whole of its lifespan

HT Low-grade ores – Ores that contain small amounts of metal

Mass number – The total number of protons and neutrons in an atom

Mixture – Two or more elements or compounds that are not chemically combined

Mole – The amount of material containing 6×10^{23} particles

Monomer – The individual molecules that join together to form a polymer

Nanometre – 1×10^{-9} (0.000 000 001) m

Nanoparticles – Particles that are less than 100 nm in size

Nanoscience – The study of structures that are 1–100 nm in size (i.e. a few hundred atoms)

Nanotube – A molecule made of carbon atoms arranged in a tubular structure

HT Ore – A naturally occurring mineral from which it is economically viable to extract a metal

HT Oxidation – A reaction involving the gain of oxygen or the loss of electrons

Particulates – Small solid particles present in the air

Peer-reviewed evidence – Work (evidence) of scientists that has been checked by other scientists to ensure that it is accurate and scientifically valid

Percentage yield – The ratio of mass of product obtained to mass of product expected

Photosynthesis – The process by which green plants and algae use water and carbon dioxide to make glucose and oxygen

HT Phytomining – A method of metal extraction that involves growing plants in metal solutions so that they accumulate metal; the plants are then burnt and the metal extracted from the ash

Polymer – A large, long-chained molecule

Potable water – Water that is safe to drink

Precipitate – A solid formed when two solutions react together

Rate – A measure of the speed of a chemical reaction

HT Redox – A reaction in which both oxidation and reduction occur

HT Reduction – A reaction involving the loss of oxygen or the gain of electrons

Relative formula mass – The sum of the atomic masses of the atoms in a formula

Salt – A product of the reaction that occurs when an acid is neutralised

Shell – Another word for an energy level

Solute – A solid that dissolves in a liquid to form a solution

Spectroscope/flame photometer – A device for observing the spectrum of light produced by a source

HT Strong acid – An acid that fully ionises when dissolved

Sustainable development – Living in a way that meets the needs of the current generation without compromising the potential of future generations to meet their own needs

Thermal decomposition – The breakdown of a chemical substance due to the action of heat

Titration – A method used to find the concentration of an acid or alkali

Unsaturated – A molecule that contains a carbon-carbon double bond

HT Weak acid – An acid that partially ionises when dissolved in water

Yield – The amount of product obtained in a reaction

The Periodic Table

Key

- Metals
- Non-metals

Relative atomic mass →		
Atomic symbol →	**H**	
Name →	hydrogen	
Atomic/proton number →	1	

1	2											3	4	5	6	7	0 or 8
																	4 **He** helium 2
7 **Li** lithium 3	9 **Be** beryllium 4											11 **B** boron 5	12 **C** carbon 6	14 **N** nitrogen 7	16 **O** oxygen 8	19 **F** fluorine 9	20 **Ne** neon 10
23 **Na** sodium 11	24 **Mg** magnesium 12											27 **Al** aluminium 13	28 **Si** silicon 14	31 **P** phosphorus 15	32 **S** sulfur 16	35.5 **Cl** chlorine 17	40 **Ar** argon 18
39 **K** potassium 19	40 **Ca** calcium 20	45 **Sc** scandium 21	48 **Ti** titanium 22	51 **V** vanadium 23	52 **Cr** chromium 24	55 **Mn** manganese 25	56 **Fe** iron 26	59 **Co** cobalt 27	59 **Ni** nickel 28	63.5 **Cu** copper 29	65 **Zn** zinc 30	70 **Ga** gallium 31	73 **Ge** germanium 32	75 **As** arsenic 33	79 **Se** selenium 34	80 **Br** bromine 35	84 **Kr** krypton 36
85 **Rb** rubidium 37	88 **Sr** strontium 38	89 **Y** yttrium 39	91 **Zr** zirconium 40	93 **Nb** niobium 41	96 **Mo** molybdenum 42	[98] **Tc** technetium 43	101 **Ru** ruthenium 44	103 **Rh** rhodium 45	106 **Pd** palladium 46	108 **Ag** silver 47	112 **Cd** cadmium 48	115 **In** indium 49	119 **Sn** tin 50	122 **Sb** antimony 51	128 **Te** tellurium 52	127 **I** iodine 53	131 **Xe** xenon 54
133 **Cs** caesium 55	137 **Ba** barium 56	139 **La*** lanthanum 57	178 **Hf** hafnium 72	181 **Ta** tantalum 73	184 **W** tungsten 74	186 **Re** rhenium 75	190 **Os** osmium 76	192 **Ir** iridium 77	195 **Pt** platinum 78	197 **Au** gold 79	201 **Hg** mercury 80	204 **Tl** thallium 81	207 **Pb** lead 82	209 **Bi** bismuth 83	[209] **Po** polonium 84	[210] **At** astatine 85	[222] **Rn** radon 86
[223] **Fr** francium 87	[226] **Ra** radium 88	[227] **Ac*** actinium 89	[261] **Rf** rutherfordium 104	[262] **Db** dubnium 105	[266] **Sg** seaborgium 106	[264] **Bh** bohrium 107	[277] **Hs** hassium 108	[268] **Mt** meitnerium 109	[271] **Ds** darmstadtium 110	[272] **Rg** roentgenium 111	[285] **Cn** copernicium 112	[286] **Uut** ununtrium 113	[289] **Fl** flerovium 114	[289] **Uup** ununpentium 115	[293] **Lv** livermorium 116	[294] **Uus** ununseptium 117	[294] **Uuo** ununoctium 118

*The lanthanides (atomic numbers 58–71) and the actinides (atomic numbers 90–103) have been omitted.

The relative atomic masses of copper and chlorine have not been rounded to the nearest whole number.

Index

Acknowledgements

The author and publisher are grateful to the copyright holders for permission to use quoted materials and images.

All images are © Shutterstock and © HarperCollins*Publishers*.

Every effort has been made to trace copyright holders and obtain their permission for the use of copyright material. The author and publisher will gladly receive information enabling them to rectify any error or omission in subsequent editions. All facts are correct at time of going to press.

Published by Collins
An imprint of HarperCollins*Publishers*
1 London Bridge Street
London SE1 9GF

HarperCollins*Publishers*
Macken House, 39/40 Mayor Street Upper, Dublin 1, D01 C9W8, Ireland

ISBN: 9780008276058

First published 2018
This edition published 2020
Previously published as Letts

10 9 8 7 6 5 4

British Library Cataloguing in Publication Data.

A CIP record of this book is available from the British Library.

Author: Dan Evans
Commissioning Editors: Clare Souza and Kerry Ferguson
Editor/Project Manager: Katie Galloway
Cover Design: Kevin Robbins
Inside Concept Design: Ian Wrigley
Text Design and Layout: Nicola Lancashire at Rose & Thorn Creative Services, and Ian Wrigley
Production: Natalia Rebow
Printed in Great Britain.

MIX
Paper | Supporting responsible forestry
FSC www.fsc.org
FSC™ C007454

This book contains FSC™ certified paper and other controlled sources to ensure responsible forest management.

For more information visit: www.harpercollins.co.uk/green